Earth Moving Operations
Manual
FM 5-434

This field manual (FM) is a guide for engineer personnel responsible for planning, designing, and constructing earthworks in the theater of operations. It gives estimated production rates, characteristics, operation techniques, and soil considerations for earthmoving equipment. This guide should be used to help select the most economical and effective equipment for each individual operation.

Should you have suggestions or feedback on ways to improve this book please send email to Books@OcotilloPress.com

Cover Photo Credit:
Flickr - New Jersey National Guard - Public Domain Photo
150th and 160th Engineer Companies, NJ Army National Guard
Photo by USAF MSGT Mark C. Olsen

Edited 2021 Ocotillo Press
ISBN 978-1-954285-36-1

Printed in the United States of America

Ocotillo Press
Houston, TX 77017
Books@OcotilloPress.com

Disclaimer: The user of this book is responsible for following safe and lawful practices at all times. The publisher assumes no responsibility for the use of the content of this book. The publisher has made an effort to ensure that the text is complete and properly typeset, however omissions, errors, and other issues may exist that the publisher is unaware of.

Field Manual
No. 5-434

*FM 5-434
Headquarters
Department of the Army
Washington, DC 15 JUNE 2000

Earthmoving Operations

Contents

Page

DISTRIBUTION RESTRICTION: Approved for public release; distribution is unlimited.

*This publication supersedes FM 5-434, 26 August 1994, and FM 5-164, 30 August 1974.

Preface

This field manual (FM) is a guide for engineer personnel responsible for planning, designing, and constructing earthworks in the theater of operations. It gives estimated production rates, characteristics, operation techniques, and soil considerations for earthmoving equipment. This guide should be used to help select the most economical and effective equipment for each individual operation.

This manual discusses the complete process of estimating equipment production rates. However, users of this manual are encouraged to use their experience and data from other projects in estimating production rates.

The material in this manual applies to all construction equipment regardless of make or model. The equipment used in this manual are examples only. Information for production calculations should be obtained from the operator and maintenance manuals for the make and model of the equipment being used.

Appendix A contains an English-to-metric measurement conversion chart.

The proponent of this publication is HQ TRADOC. Send comments and recommendations on Department of the Army (DA) Form 2028 directly to United States Army Engineer School (USAES), ATTN: ATSE-DOT-DD, Directorate of Training, 320 Engineer Loop Suite 336, Fort Leonard Wood, Missouri 65473-8929.

Unless this publication states otherwise, masculine nouns and pronouns do not refer exclusively to men.

Chapter 1

Managing Earthmoving Operations

Earthmoving may include site preparation; excavation; embankment construction; backfilling; dredging; preparing base course, subbase, and subgrade; compaction; and road surfacing. The types of equipment used and the environmental conditions will affect the man- and machine-hours required to complete a given amount of work. Before preparing estimates, choose the best method of operation and the type of equipment to use. Each piece of equipment is specifically designed to perform certain mechanical tasks. Therefore, base the equipment selection on efficient operation and availability.

PROJECT MANAGEMENT

1-1. Project managers must follow basic management phases to ensure that construction projects successfully meet deadlines set forth in project directives. Additionally, managers must ensure conformance to safety and environmental-protection standards. The basic management phases as discussed in FM 5-412 are—

- Planning.
- Organizing.
- Staffing.
- Directing.
- Controlling.
- Executing.

EQUIPMENT SELECTION

1-2. Proper equipment selection is crucial to achieving efficient earthmoving and construction operations. Consider the machine's operational capabilities and equipment availability when selecting a machine for a particular task. The manager should visualize how best to employ the available equipment based on soil considerations, zone of operation, and project-specific requirements. Equipment production-estimating procedures discussed in this manual help quantify equipment productivity.

PRODUCTION ESTIMATES

1-3. Production estimates, production control, and production records are the basis for management decisions. Therefore, it is helpful to have a common method of recording, directing, and reporting production. (Refer to specific, equipment production-estimating procedures in the appropriate chapters in this manual.)

PRODUCTION-RATE FORMULA

1-4. The most convenient and useful *unit of work done* and *unit of time* to use in calculating productivity for a particular piece of equipment or a particular job is a function of the specific work-task being analyzed. To make accurate and meaningful comparisons and conclusions about production, it is best to use standardized terms.

$$\text{Production rate} = \frac{\text{unit of work done}}{\text{unit of time}}$$

- **Production rate.** The entire expression is a time-related production rate. It can be cubic yards per hour, tons per shift (also indicate the duration of the shift), or feet of ditch per hour.
- **Unit of work done.** This denotes the unit of production accomplished. It can be the volume or weight of the material moved, the number of pieces of material cut, the distance traveled, or any similar measurement of production.
- **Unit of time.** This denotes an arbitrary time unit such as a minute, an hour, a 10-hour shift, a day, or any other convenient duration in which the unit of work done is accomplished.

TIME-REQUIRED FORMULA

1-5. The inverse of the production-rate formula is sometimes useful when scheduling a project because it defines the time required to accomplish an arbitrary amount of work.

$$\text{Time required} = \frac{\text{unit of time}}{\text{unit of work done}}$$

NOTE: Express the time required in units such as hours per 1,000 cubic yards, hours per acre, days per acre, or minutes per foot of ditch.

MATERIAL CONSIDERATIONS

1-6. Depending on where a material is considered in the construction process, during excavation versus after compaction, the same material weight will occupy different volumes *(Figure 1-1)*. Material volume can be measured in one of three states:

- **Bank cubic yard (BCY).** A BCY is 1 cubic yard of material as it lies in its natural/undisturbed state.
- **Loose cubic yard (LCY).** A LCY is 1 cubic yard of material after it has been disturbed by an excavation process.
- **Compacted cubic yard (CCY).** A CCY is 1 cubic yard of material after compaction.

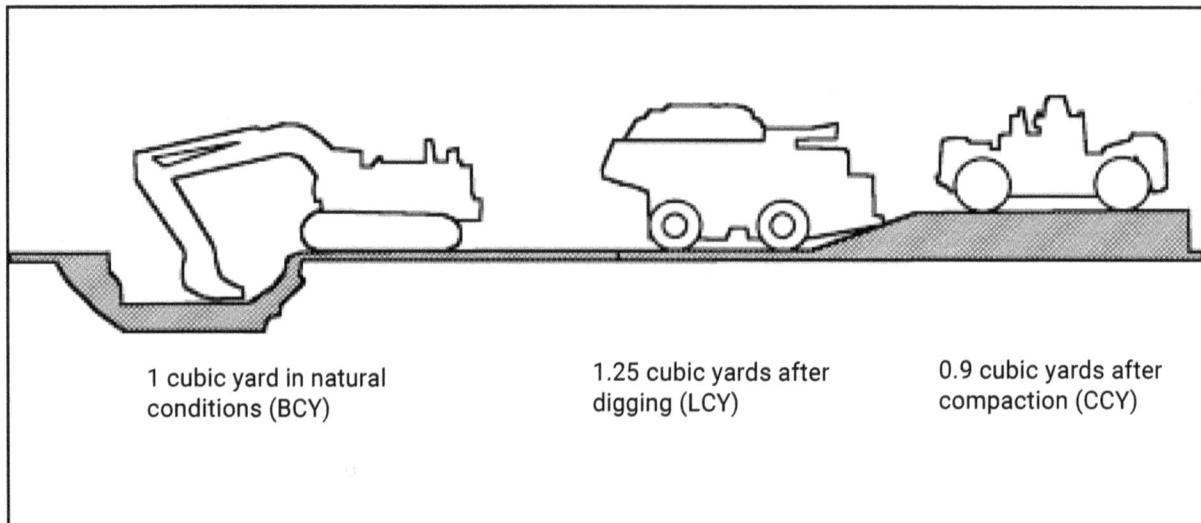

1 cubic yard in natural conditions (BCY)

1.25 cubic yards after digging (LCY)

0.9 cubic yards after compaction (CCY)

Figure 1-1. Material-Volume Changes Caused by Construction Processes

1-7. When manipulating the material in the construction process, its volume changes. *(Tables 1-1* and *1-2, page 1-4,* give material-volume conversion and load factors.) The prime question for an earthmover is about the nature of the material's physical properties; for example, how easy is it to move? For earthmoving operations, material is placed in three categories—rock, soil (common earth), and unclassified.

- **Rock.** Rock is a material that ordinary earthmoving equipment cannot remove. Fracturing rock requires drilling and blasting. After blasting, use excavators to load the rock fragments into haul units for removal.
- **Soil.** Soils are classified by particle-size distribution and cohesiveness. For instance, gravel and sands have blocky-shaped particles and are noncohesive, while clay has small, platy-shaped particles and is cohesive. Although ripping equipment may be necessary to loosen consolidated deposits, soil removal does not require using explosives.
- **Unclassified.** The unclassified (rock-soil) combination is the most common material found throughout the world. It is a mixture of rock and soil materials.

SOIL PROPERTIES

1-8. In an earthmoving operation, thoroughly analyze the material's properties (loadability, moisture content, percentage of swell, and compactability) and incorporate this information into the construction plan. Soil preparation and compaction requirements are discussed in *Chapter 11.*

Loadability

1-9. Loadability is a general material property or characteristic. If the material is easy to dig and load, it has high loadability. Conversely, if the material is difficult to dig and load, it has low loadability. Certain types of clay and loam are easy to doze or load into a scraper from their natural state.

Moisture Content

1-10. Moisture content is a very important factor in earthmoving work since moisture affects a soil's unit weight and handling properties. All soil in its natural state contains some moisture. The amount of moisture retained depends on the weather, the drainage, and the soil's retention properties. Mechanical or chemical treatment can sometimes change the moisture content of a soil. Refer to *Chapter 11* for information about increasing and decreasing the soil's moisture content.

Table 1-1. Material Volume Conversion Factors

Material Type	Converted From	Converted To		
		Bank (In Place)	Loose	Compacted
Sand or gravel	Bank (in place)	—	1.11	0.95
	Loose	0.90	—	0.86
	Compacted	1.05	1.17	—
Loam (common earth)	Bank (in place)	—	1.25	0.90
	Loose	0.80	—	0.72
	Compacted	1.11	1.39	—
Clay	Bank (in place)	—	1.43	0.90
	Loose	0.70	—	0.63
	Compacted	1.11	1.59	—
Rock (blasted)	Bank (in place)	—	1.50	1.30
	Loose	0.67	—	0.87
	Compacted	0.77	1.15	—
Coral (comparable to lime rock)	Bank (in place)	—	1.50	1.30
	Loose	0.67	—	0.87
	Compacted	0.77	1.15	—

Table 1-2. Material Weight, Swell Percentages, and Load Factors

Material Type	Loose (Pounds Per Cubic Yards)	Swell (Percent)	Load Factor	Bank (Pounds Per Cubic Yard)
Cinders	800 to 1,200	40 to 55	0.65 to 0.72	1,100 to 1,860
Clay, dry	1,700 to 2,000	40	0.72	2,360 to 2,780
Clay, wet	2,400 to 3,000	40	0.72	3,360 to 4,200
Earth (loam or silt), dry	1,900 to 2,200	15 to 35	0.74 to 0.87	2,180 to 2,980
Earth (loam or silt), wet	2,800 to 3,200	25	0.80	3,500 to 4,000
Gravel, dry	2,700 to 3,000	10 to 15	0.87 to 0.91	2,980 to 3,450
Gravel, wet	2,800 to 3,100	10 to 15	0.87 to 0.91	3,080 to 3,560
Sand, dry	2,600 to 2,900	10 to 15	0.87 to 0.91	2,860 to 3,340
Sand, wet	2,800 to 3,100	10 to 15	0.87 to 0.91	3,080 to 3,560
Shale (soft rock)	2,400 to 2,700	65	0.60	4,000 to 4,500
Trap rock	2,700 to 3,500	50	0.66	4,100 to 5,300

NOTE: The above numbers are averages for common materials. Weights and load factors vary with such factors as grain size, moisture content, and degree of compaction. If an exact weight for a specific material must be determined, run a test on a sample of that particular material.

Percentage of Swell

1-11. Most earth and rock materials swell when removed from their natural resting place. The volume expands because of voids created during the excavation process. After establishing the general classification of a soil, estimate the percentage of swell. Express swell as a percentage increase in volume (*Table 1-2*). For example, the swell of dry clay is 40 percent, which means that 1 cubic yard of clay in the bank state will fill a space of 1.4 cubic yards in a loosened state. Estimate the swell of a soil by referring to a table of material properties such as *Table 1-2*.

Compactability

1-12. In earthmoving work, it is common to compact soil to a higher density than it was in its natural state. This is because there is a correlation between higher density and increased strength, reduced settlement, improved bearing capacity, and lower permeability. The project specifications will state the density requirements.

SOIL WEIGHT

1-13. Soil weight affects the performance of the equipment. To estimate the equipment requirements of a job accurately, the unit weight of the material being moved must be known. Soil weight affects how dozers push, graders cast, and scrapers load the material. Assume that the volumetric capacity of a scraper is 25 cubic yards and that it has a rated load capacity of 50,000 pounds. If the material being carried is relatively light (such as cinder), the load will exceed the volumetric capacity of the scraper before reaching the gravimetric capacity. Conversely, if the load is gravel (which may weigh more than 3,000 pounds per cubic yard), it will exceed the gravimetric capacity before reaching the volumetric capacity. See *Table 1-2* for the unit weight of specific materials.

NOTE: The same material weight will occupy different volumes in BCY, LCY, and CCY. In an earthmoving operation, the basic unit of comparison is usually BCY. Also, consider the material in its loose state (the volume of the load). *Table 1-1* gives average material conversion factors for earth-volume changes.

LOAD FACTOR

1-14. Use a load factor (see *Table 1-2)* to convert the volume of LCY measured to BCY measured **LCY ×load factor** = **BCY(** Use similar factors) when converting material to a compacted state. The factors depend on the degree of compaction. Compute the load factor as follows:

If 1 cubic yard of clay (bank state) = 1.4 cubic yards of clay (loose state),

then 1 cubic yard of clay (loose state) $= \frac{1}{1.4}$ **or 0.72 cubic yard of clay (bank state).**

In this case, the load factor for dry clay is 0.72. This means that if a scraper is carrying 25 LCY of dry clay, it is carrying 18 BCY **(25 x 0.72)**.

ZONES OF OPERATION

1-15. The relationship of specific zones of operation to various types of earthmoving equipment is significant when selecting earthmoving equipment. A mass diagram graphically depicts how materials should be moved and is a good tool for determining the zones of operation. Mass diagrams are explained in FM 5-430-00-1. There are three zones of operation to consider on a construction project.

POWER ZONE

1-16. In the power zone, maximum power is required to overcome adverse site or job conditions. Such conditions include rough terrain, steep slopes, pioneer operations, or extremely heavy loads. The work in these areas requires crawler tractors that can develop high drawbar pull at slow speeds. In these adverse conditions, the more traction a tractor develops, the more likely it will reach its full potential.

SLOW-SPEED HAULING ZONE

1-17. The slow-speed hauling zone is similar to the power zone since power, more than speed, is the essential factor. Site conditions are slightly better than in the power zone, and the haul distance is short. Since improved conditions give the dozer more power, and distances are too short for most scrapers to build up sufficient momentum to shift into higher speeds, both machines achieve the same speed. Considerations that determine a slow-speed hauling zone are as follows:

- The ground conditions do not permit rapid travel and the movement distance of the material is beyond economical dozing operations.
- The haul distances are not long enough to permit scrapers to travel at high speeds.

HIGH-SPEED HAULING ZONE

1-18. In the high-speed hauling zone, construction has progressed to where ground conditions are good, or where long, well-maintained haul roads are established. Achieve this condition as soon as possible. Production increases when the scraper is working at its maximum speed. Considerations that determine a high-speed hauling zone are as follows:

- Good hauling conditions exist on both grade and haul-road surfaces.
- Haul distances are long enough to permit acceleration to maximum travel speeds.
- Push tractors (also referred to as pushers) are available to assist in loading.

CAUTION
Operate equipment at safe speeds to prevent personal injury or premature failure of the machine's major components. Accomplish hauling operations safely as well as efficiently.

Chapter 2

Dozers

Dozers (tracklaying crawlers or wheel tractors equipped with a blade) are perhaps the most basic and versatile items of equipment in the construction industry. Dozers are designed to provide high drawbar pull and traction effort. They are the standard equipment for land clearing, dozing, and assisting in scraper loading. They can be equipped with rear-mounted winches or rippers. Crawler tractors exert low ground-bearing pressure, which adds to their versatility. For long moves between projects or within a project, transport dozers on heavy trailers. Moving them under their own power, even at slow speeds, increases track wear and shortens the machine's operational life.

DESCRIPTION

2-1. A crawler dozer consists of a power plant (typically a diesel engine) mounted on an undercarriage, which rides on tracks. The tracks extend the full length of the dozer. There are two classifications of military dozers, based on weight and pounds of drawbar pull. The light class (about 16,000 pounds operating weight) includes the deployable universal combat earthmover (DEUCE) *(Figure 2-1)*. The medium class includes dozers having an operating weight of 15,000 to 45,000 pounds *(Figure 2-2, page 2-2)*.

Figure 2-1. DEUCE, Light-Class Dozer

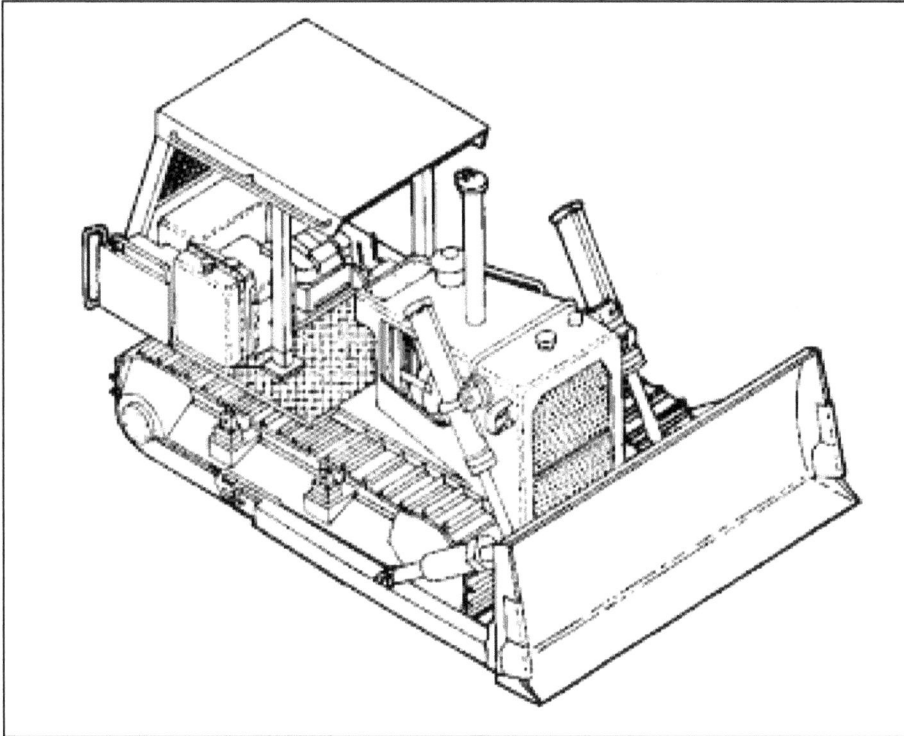

Figure 2-2. Medium-Class Dozer

BLADES

2-2. A dozer blade consists of a moldboard with replaceable cutting edges and side bits. Either the push arms and tilt cylinders or a C-frame are used to connect the blade to the tractor. Blades vary in size and design based on specific work applications. The hardened-steel cutting edges and side bits are bolted on because they receive most of the abrasion and wear out rapidly. This allows for easy replacement. Machine designs allow either edge of the blade to be raised or lowered in the vertical plane of the blade (tilt). The top of the blade can be pitched forward or backward varying the angle of attack of the cutting edge (pitch). Blades mounted on a C-frame can be turned from the direction of travel (angling). These features are not applicable to all blades, but any two of these features may be incorporated in a single mount.

STRAIGHT BLADE

2-3. Use straight blades for pushing material and cutting ditches. This blade is mounted in a fixed position, perpendicular to the line of travel. It can be tilted and pitched either forward or backward within a 10° arc. Tilting the blade allows concentration of dozer driving power on a limited length of the blade. Pitching the blade provides increased penetration for cutting or less penetration for back dragging.

ANGLE BLADE

2-4. Angle blades, which are 1 to 2 feet wider than straight blades, are used most effectively to side cast material when backfilling or when making sidehill cuts. Use an angle blade for rough grading, spreading piles, or windrowing material. It can be angled up to a maximum of 25° left or right of perpendicular to the dozer or used as a straight blade. When angled, the blade can be tilted but it cannot be pitched.

SPECIAL-PURPOSE BLADE

2-5. There are special blades *(Figure 2-3),* such as the Rome K/G, designed for clearing brush and trees but not for earthmoving. The Rome K/G blade is permanently fixed at an angle. On one end of the blade is a stinger. This stinger consists of a vertical splitter and stiffener and a triangular-shaped horizontal part called the web. One side of the triangular web abuts the bottom of the vertical splitter, and the other side abuts the cutting edge of the blade. The abutting sides of the web are each about 2 feet in length, depending on how far the stinger protrudes from the blade. This blade is designed to cut down brush and trees at, or a few inches above, ground level rather than uprooting them. When cutting a large-diameter tree, first use the stinger to split the tree to weaken it; then, cut the tree off and push it over with the blade. Keep both the stinger and the cutting edge sharp. The operator must be well-trained to be efficient in this operation. There are other special-purpose blades not discussed in this manual which can be mounted on dozers.

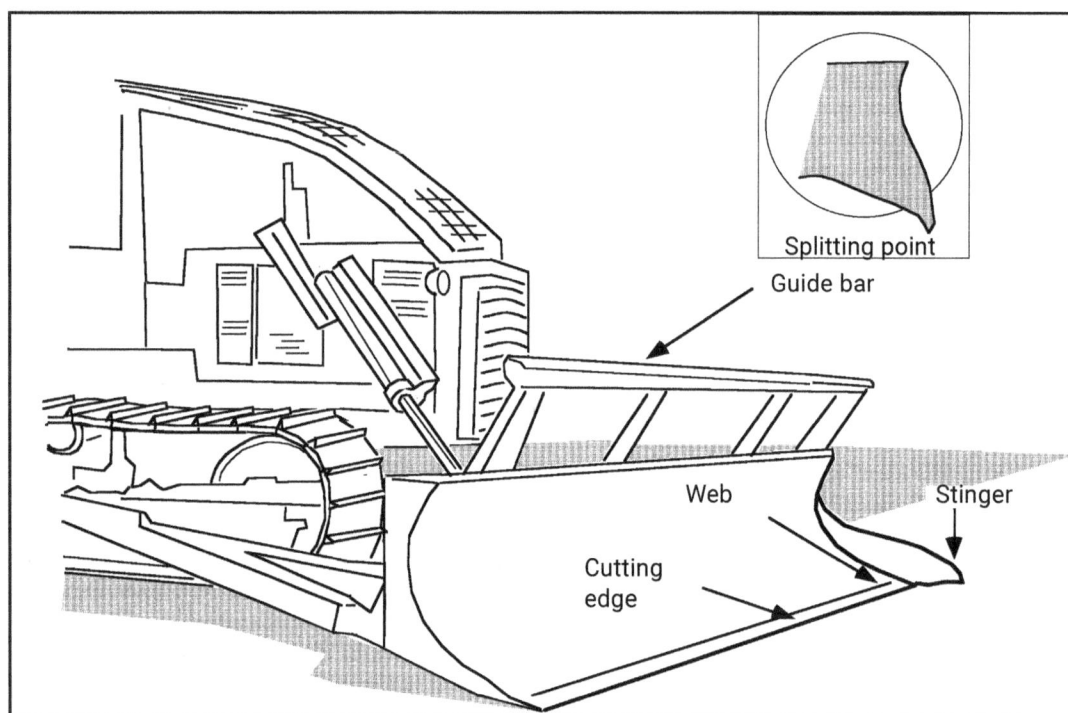

Figure 2-3. Special-Purpose Clearing Blade

CLEARING AND GRUBBING OPERATIONS

2-6. Clearing vegetation and trees is usually necessary before moving and shaping the ground. Clearing includes removing surface boulders and other materials embedded in the ground and then disposing of the cleared material. Ensure that environmental-protection considerations are addressed before conducting clearing operations. Specifications may allow shearing of the vegetation and trees at ground level, or it may be necessary to grub (removing stumps and roots from below the ground). Project specifications will dictate the proper clearing techniques. Plan clearing operations to allow disposal of debris in one handling. It is best to travel in one direction when clearing. Changing direction tends to skin and scrape the trees instead of uprooting them or allowing a clean cut. Clearing techniques vary with the type of vegetation being cleared, the ground's soil type, and the soil's moisture condition. *Table 2-1* shows average clearing rates for normal area-clearing jobs. Increase the *Table 2-1* values by 60 percent if the project requires strip-type clearing (common in tactical land clearing). Engineers perform tactical land clearing as a combat support function intended to enhance and complement mobility, firepower, surveillance, and target acquisition.

Table 2-1. Quick Production Estimates for Normal Area Clearing

Equipment	Equipment (Hours Per Acre)		
	Light (12 Inches or Less*)	Medium (12 to 18 Inches*)	Heavy (18 Inches*)
Bulldozer:			
Medium tractor	2.5	5.0	10.0
Heavy tractor	1.5	3.0	8.0
Shear blade:			
Medium tractor	0.4	0.8	1.3
Heavy tractor	0.3	0.5	0.8
*Maximum tree size			
NOTE: These clearing rates are average for tree counts of 50 trees per acre. Adverse conditions (slopes, rocks, soft ground) can reduce these rates significantly.			

BRUSH AND SMALL TREES

2-7. Moving the dozer, with the blade slightly below ground level, will usually remove small trees and brush. The blade cuts, breaks off, or uproots most of the tree and bends the rest for removal on the return trip. A medium tractor with a dozer blade can clear and pile about 0.25 acres of brush or small trees per hour.

MEDIUM TREES

2-8. To remove a medium-size tree (7 to 12 inches in diameter), raise the blade as high as possible to gain added leverage and then push the tree over slowly. As the tree starts to fall, back the dozer quickly to avoid the rising roots. Then lower the blade and drive the dozer forward, lifting out the roots. The average time for a medium tractor with a dozer blade to clear and pile medium trees is 2 to 9 minutes per tree.

LARGE TREES

2-9. Removing large trees (12 to 30 inches in diameter) is much slower and more difficult than clearing brush and smaller trees. First, gently and cautiously probe the tree for dead limbs that could fall. Determine the tree's natural direction of lean, if any; this is the best direction for pushing the tree over. Then, position the blade high and center it on the tree for maximum leverage. If possible, push the tree over the same as a medium tree. However, if the tree has a massive, deeply embedded root system, use the following method *(Figure 2-4):*

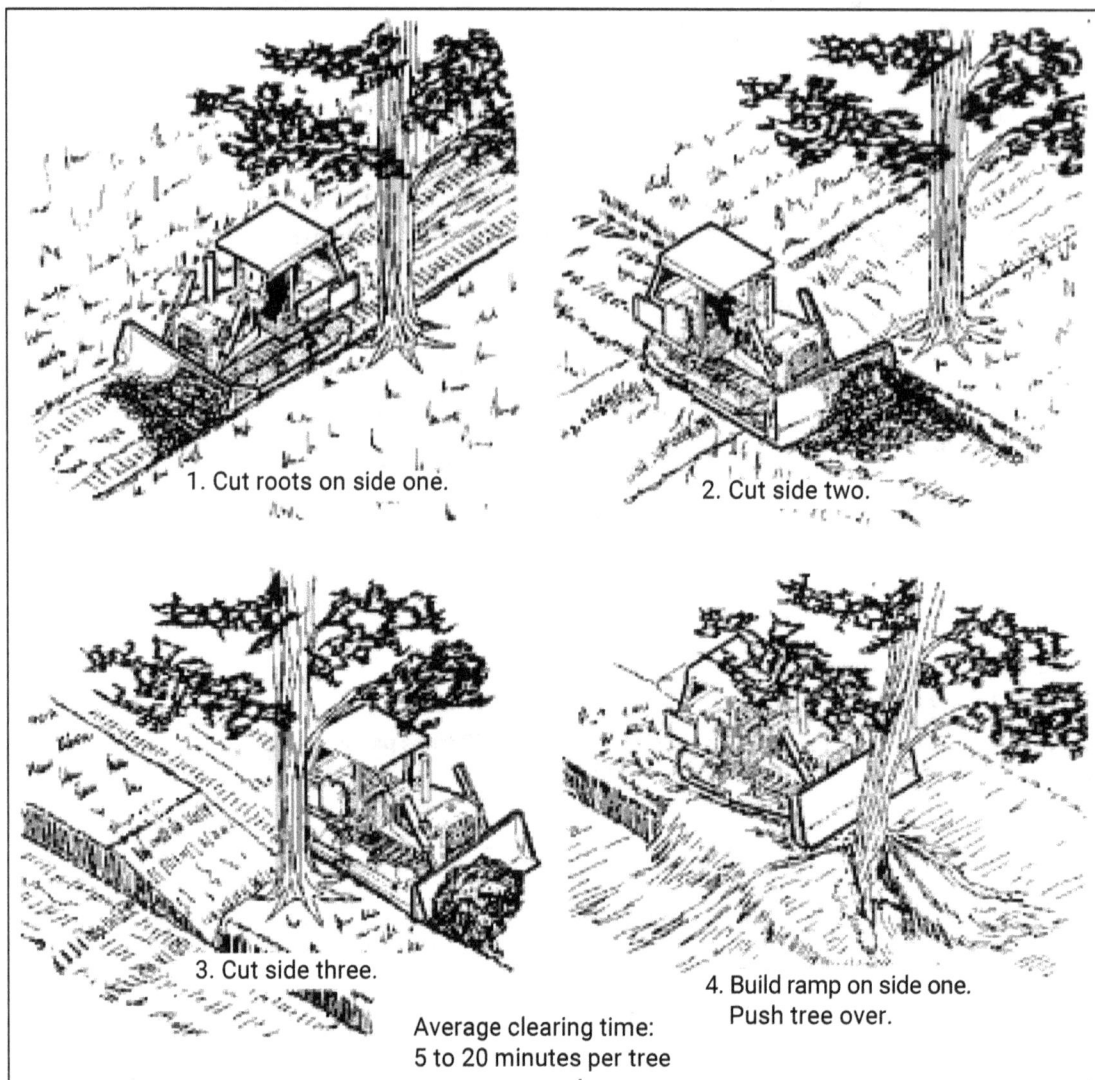

Figure 2-4. Four Steps for Removing a Large Tree With a Massive, Deeply Embedded Root System

Step 1. Start on the side opposite the proposed direction of fall, and make a cut deep enough to sever some of the large roots. Make the cut like a V-ditch, tilted downward laterally toward the roots.

Step 2. Cut side two.

Step 3. Cut side three.

Step 4. Build an earth ramp on the same side as the original cut to obtain greater pushing leverage. Then push the tree over and, as the tree starts to fall, reverse the dozer quickly to avoid the rising root mass. After felling the tree, fill the stump hole so that it will not collect water.

The average time for a medium tractor with a dozer blade to clear and pile large trees is 5 to 20 minutes per tree. The time required to clear and pile massive trees requiring this four-step procedure will often be more than 20 minutes each.

NOTE: The roots on the fourth side may also need to be cut.

ROOTS

2-10. Mount a rake on the dozer in place of the blade to remove roots and small stumps. As the dozer moves forward, it forces the teeth of the rake below the ground's surface. The teeth will catch the belowground roots and the surface brush left from the felling operation, while the soil remains or passes through.

SAFETY PRECAUTIONS

2-11. Never operate clearing tractors too close together. Do not follow a tree too closely when pushing it, because when it begins to fall, its stump and roots may catch under the front of the dozer. Clean out accumulated debris in the dozer's belly pan often to prevent fires in the engine compartment.

PRODUCTION ESTIMATES

2-12. The two methods for estimating production for clearing and grubbing projects are the quick method and the tree-count method.

Quick Method

2-13. *Table 2-1, page 2-4,* shows quick estimates for normal area clearing. Use the quick method only when a detailed reconnaissance and a tree count are not possible.

Step 1. Determine the size of the area to clear (in acres).

$$\text{Acres to be cleared} = \frac{\text{width (feet)} \times \text{length (feet)}}{43,560 \text{ square feet per acre}}$$

Step 2. Determine the size and number of dozers available.

Step 3. Determine the maximum size of the trees to clear.

Step 4. Determine the time required (hours per acre) for clearing, based on dozer size and tree size (see *Table 2-1*).

Step 5. Determine the efficiency factor for the work. Operators require breaks, and there are always secondary delays for minor equipment repairs. Therefore, actual production time per hour is something less than 60 minutes. In the case of a well-managed job, expect 50 minutes of production time per hour.

$$\text{Efficiency factor} = \frac{\text{actual working minutes per hour}}{60\text{-minute working hour}}$$

Step 6. Determine the operator factor using *Table 2-2*.

Step 7. Determine the total time (in hours) required to complete the mission.

$$\text{Total time (hours)} = \frac{D \times A}{E \times O \times N}$$

where—
  D = time required, in hours per acre
  A = total area, in acres
  E = efficiency factor
  O = operator factor
  N = number of dozers available

Table 2-2. Operator Factors for Track Dozers

Operator Ability	Daylight	Night
Excellent	1.00	0.75
Average	0.75	0.56
Poor	0.60	0.45

NOTE: These factors assume good visibility and a 60-minute working hour efficiency.

EXAMPLE

Determine the time required to clear an area that is 500-feet wide by 0.5 mile long. Two medium bulldozers are available for the task. The largest trees in the area are 14 inches in diameter, and the ground is fairly level. The operators are of average ability and will do all work during daylight hours. Expected efficiency is 50 minutes per hour.

Step 1. Total area in acres $= \dfrac{\text{width (feet)} \times \text{length (feet)}}{43{,}560 \text{ square feet per acre}}$

$= \dfrac{500 \text{ feet} \times (0.5 \text{ mile} \times 5{,}280 \text{ feet per mile})}{43{,}560} = 30.3 \text{ acres}$

Step 2. Dozer size = medium
    Number of dozers available = 2

Step 3. Maximum tree size = 14 inches

Step 4. Time required = 5 hours per acre *(Table 2-1, page 2-4)*

Step 5. Efficiency factor $= \dfrac{50 \text{ minutes per hour}}{60\text{-minute working hour}} = 0.83$

Step 6. Operator factor = 0.75 *(Table 2-2)*

Step 7. Total time (hours) $= \dfrac{5 \text{ hours per acre} \times 30.3 \text{ acres}}{0.83 \times 0.75 \times 2} = 121.6 \text{ or } 122 \text{ hours}$

Tree-Count Method

2-14. Use this method when a detailed reconnaissance and a tree count are possible. The tree-count method allows for a better production estimate.

Step 1. Determine the size of the area to clear (in acres). Refer to step 1 of the quick method.

Step 2. Determine the size and number of dozers available.

Step 3. Determine the average number of each size of tree per acre. This will require a field reconnaissance.

Step 4. Determine the basic production factors (hours per acre) based on the dozer size and the size of the trees to clear *(Table 2-3)*.

Table 2-3. Production Factors for Felling With a Clearing Blade

Tractor	Base Minutes Per Acre B	Tree Diameter Range				
		1-2 Feet M_1	2-3 Feet M_2	3-4 Feet M_3	4-6 Feet M_4	More Than 6 Feet F
Medium	23.48	0.5	1.7	3.6	10.2	3.3
Heavy	18.22	0.2	1.3	2.2	6.0	1.8
NOTE: These times are based on working on reasonably level ground with good footing and an average mix of soft and hardwoods.						

Step 5. Determine the time required to clear one acre.

$$D = H([\ A \times\ ]B + [M_1 \times N_1] + [M_2 \times N_2] + [M_3 \times N_3]\ [M_4 \times N_4]\ +[I \times F])$$

where—

D = clearing time of one acre, in minutes

H = hardwood factor affecting total time—

    H = 1.3 if hardwoods are 75 to 100 percent

    H = 1 if hardwoods are 25 to 75 percent

    H = 0.7 if hardwoods are 0 to 25 percent

A = tree-density and presence-of-vines factor affecting total time

    A = 2 if density is more than 600 trees per acre (dense)

    A = 1 if density is 400 to 600 trees per acre (medium)

    A = 0.7 if density is less than 400 trees per acre (light)

    A = 2 if heavy vines are present

B = base time per acre determined from dozer size, in minutes

M = time required per tree in each diameter range, in minutes

N = number of trees per acre in each diameter range, from reconnaissance

I = sum of diameter of all trees per acre greater than 6 feet in diameter at ground level (in foot increments), from reconnaissance

F = time required per foot of diameter for trees greater than 6 feet in diameter, in minutes

NOTE: When it is necessary to grub roots and stumps, increase the time per acre by 25 percent.

Step 6. Determine the total time (in hours) required to complete the mission.

$$\text{Total time (hours)} = \frac{D \times A}{N}$$

where—

  D = time required to clear one acre (from step 5), in hours
  A = total area
  N = number of dozers

NOTE: The tree-count method has no correction factor for efficiency or operator skill. The values in *Table 2-3* are based on normal efficiency and average operator skill.

SIDEHILL EXCAVATIONS

2-15. One of a dozer's more important uses is making sidehill cuts, which includes pioneering road cuts along hillsides. An angle blade is preferred for this operation because of its side-casting ability.

CREATING A SLOPE

2-16. It is best to start the cut at the top of the hill, creating a bench several dozer lengths long. Do this by working up and down the slope perpendicular to the long direction of the project *(Figure 2-5[A], page 2-10.)* Design the benches to ensure that water runs off without damaging the slope. If possible, start the bench on the uphill extreme of the cut (the highest point of the cut) and then widen and deepen the cut until the desired road profile is achieved. Be sure to start the bench far enough up the slope to allow room for both the inner slope and the roadway.

NOTE: When working on extremely steep slopes, a winch line may be necessary to stabilize the dozer (see *paragraph 2-37).*

2-17. Because the perpendicular passes are short, the dozer usually is not able to develop a full blade load. Therefore, after constructing the initial bench, turn the dozer and work in the long direction of the project *(Figure 2-5[B], page 2-10).* Develop a full blade load and then turn the dozer to push the material over the side. After developing the bench, use either a dozer or a scraper to complete the cut. Keep the inside (hillside) of the roadway lower than the outside. This allows the dozer to work effectively on the edge and decreases the erosion of the outer slope. Make sure to maintain the proper slope on the inside of the cut. It is very difficult to change the cut slope after construction. Maintain the proper bench slope by moving out from the inside slope on each successive cut. Determine the slope ratio from the distance moved away from the slope for each successive cut and the depth of each cut. When cutting the road's cross slope, work from the toe of the bench to the road's outside edge.

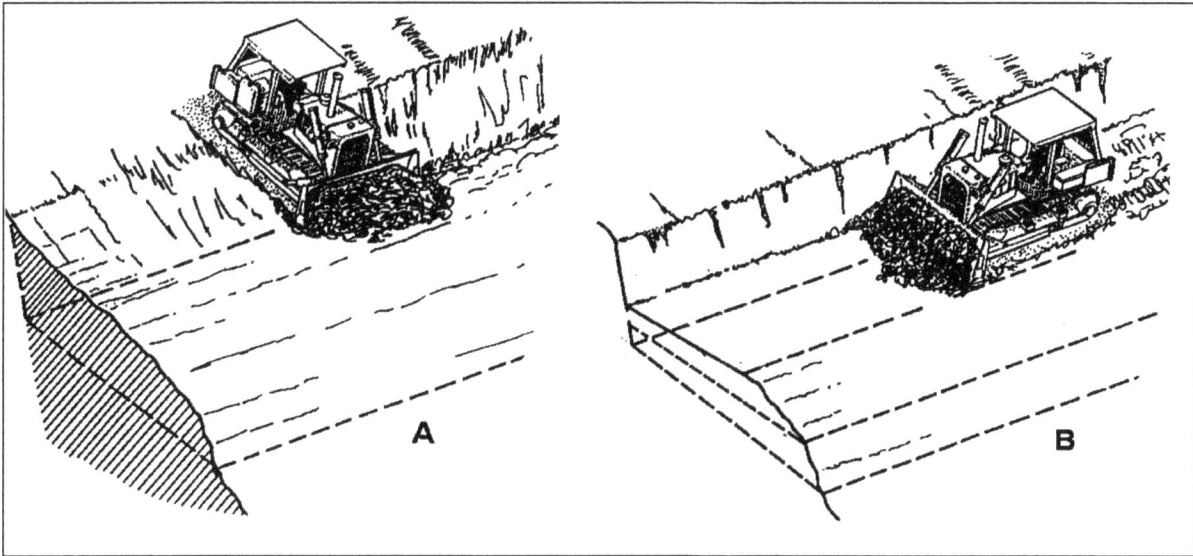

Figure 2-5. Sidehill Cut

FINISHING A SIDE SLOPE

2-18. There are two methods for finishing a side slope—working perpendicular to the slope and working diagonally up the slope.

Working Perpendicular to the Slope

2-19. The dozer shown in *Figure 2-6* is finishing a side slope by working perpendicular to the slope. Start the dozer at the top of the embankment and, on each pass, earth will fall to the lower side of the blade forming a windrow. On succeeding passes, pick up this windrow and use it to fill holes and other irregularities in the terrain. Be careful to prevent the blade corner from digging in too deep; this would steepen the slope beyond job specifications.

Figure 2-6. Finishing a Side Slope Working Perpendicular to the Slope

Working Diagonally Up the Slope

2-20. The dozer shown in *Figure 2-7* is finishing the side slope by starting at the bottom and working diagonally up the slope. The windrow that forms is continually pushed to one side, which tends to fill low spots, holes, and irregularities. This is one of the few instances where a dozer works effectively pushing uphill.

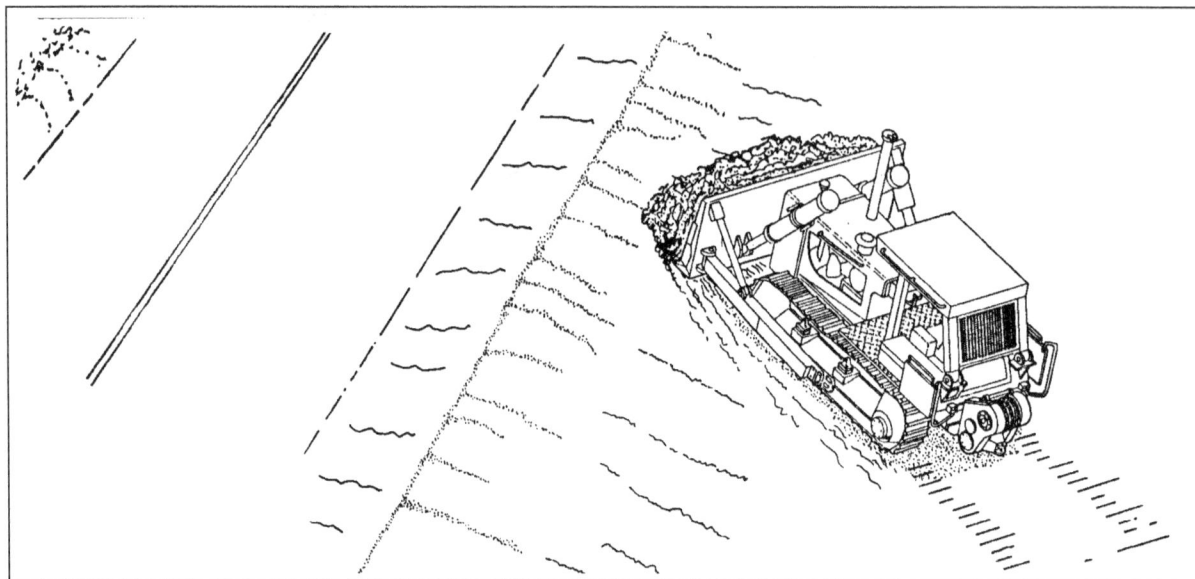

Figure 2-7. Finishing a Side Slope Working Diagonally up the Slope

OPERATION TECHNIQUES

2-21. Dozers work best when the ground is firm and without potholes, sharp ridges, or rocks. Uneven surfaces make it difficult to keep the blade in contact with the ground. This tends to bury vegetation in hollows rather than remove it. To save time and increase output, use the following techniques when conditions permit.

DOZING

2-22. When straight dozing, if the blade digs in and the rear of the machine rises, raise the blade to continue an even cut. If moving a heavy load causes the travel speed to drop, shift to a lower gear and/or raise the blade slightly. When finishing or leveling, a full blade handles easier than a partially-loaded blade.

Side-by-Side Dozing

2-23. Side-by-side dozing will increase production 15 to 25 percent when moving material 50 to 300 feet *(Figure 2-8, page 2-12)*. When the distance is less than 50 feet, the extra time needed to maneuver and position the dozers will offset the increased production.

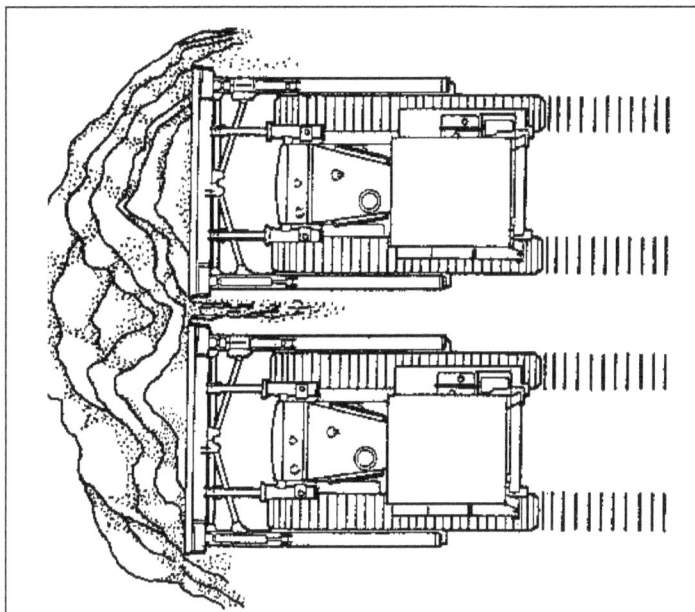

Figure 2-8. Side-by-Side Dozing

Slot Dozing

2-24. Slot dozing uses spillage from the first few passes to build a windrow on each side of a dozer's path *(Figure 2-9)*. This forms a trench, preventing blade-side spillage on subsequent passes. To increase production, align cuts parallel, leaving a narrow uncut section between slots. Then, remove the uncut section by normal dozing. When grade and soil conditions are favorable, slot dozing can increase output by as much as 20 percent.

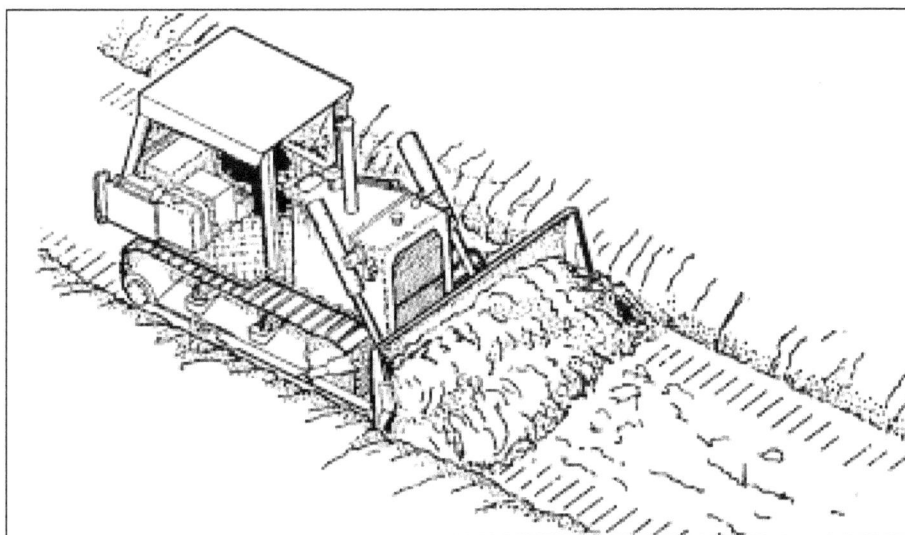

Figure 2-9. Slot Dozing

Downhill Dozing

2-25. Pile several loads at the brink of the hill, and then push them to the bottom in one pass. When dozing downhill, travel to the bottom of the hill with each load. Use downhill dozing whenever possible since it increases production.

Hard-Materials Dozing

2-26. Use the dozer blade to loosen hard material when rippers are not available. Tilt the blade to force one corner into the material. Tilting is done through blade control, by driving one track onto a ridge of material bladed up for this purpose or by placing a rock or log under the track. To maximize the driving force of the blade, hook only the tilted end under the material. Break a thin layer by turning on it with a dozer. Turning causes the track grousers (cleats) to break through the top layer. With a thin layer of frozen material, it is best to break through at one point. By lifting and pushing, the blade breaks through the top frozen layer as shown in *Figure 2-10.*

Figure 2-10. Dozing Hard Materials or Frozen Ground Layers

Rock Dozing

2-27. Use a rake to remove small rocks. The rake lets the soil remain, or pass through, while digging the rocks from the earth. When removing large, partially buried boulders, tilt the dozer blade and dig the earth out from around three sides of the boulder. Lower the blade enough to get under the fourth side. Lift the blade as the dozer moves forward to create a lifting, rolling action of the boulder. If the dozer cannot push the boulder, lift it upward with the blade and have someone place a log or some other object under the boulder so the dozer can get another hold. The rolling action removes the boulder as the dozer moves forward. Dozer work in rocky areas increases track wear. If possible, install rock shoes or rock pads to cut down on this wear.

Wet-Materials Dozing

2-28. Wet material is difficult to move with a dozer. Also, the wet ground may be too soft to support the weight of the dozer. If so, make each successive pass the full depth of the wet material. This will place the dozer on a firmer footing. If available, use wider tracked shoes for better flotation. When working in mud, push the mud back far enough that it will not flow back into the cut. Make provisions for recovery operations in case the dozer becomes stuck. Try to use machines equipped with a winch.

DITCHING

2-29. Shallow ditches are best accomplished using a grader, but dozers can accomplish rough ditching. Tilt the dozer blade to cut shallow V-ditches *(Figure 2-11)*. For larger ditches, push the material perpendicular to the center line of the ditch. After reaching the desired depth, push the material the length of the ditch to smooth the sides and bottom. Many times it is necessary to correct irregularities in a ditch. Attempt to remove humps or fill holes in a single pass. Use multiple passes to correct the grade.

Figure 2-11. Tilt Dozer Ditching

CONSTRUCTING A STOCKPILE

2-30. A dozer is a good machine for creating stockpiles of material that can then be easily loaded into haul units by either a loader or a hydraulic hoe excavator. Use the following steps to construct a stockpile:

Step 1. Push the material from the beginning of the excavation to the stockpile area on the first pass. This distance should be no more than 75 feet from the start point. Do not excavate deeper than 6 to 8 inches, while maintaining a smooth cut.

WARNING

Before putting the machine in reverse, and while backing, the operator must be satisfied that no one will be endangered.

Step 2. Begin to raise the blade one dozer length from the stockpile, letting the material drift under the blade forming a ramp upon reaching the stockpile area.

CAUTION

Keep the dozer under control at all times. Do not put the transmission into neutral to allow the machine to coast. Select the gear range necessary before starting down the grade. Do not change gears while going downhill.

Step 3. Push the material on successive cuts in the same manner, working the dozer from the start point all the way around the work area while stockpiling. Overlap cuts about one-third of the blade's width to pick up windrows.

NOTE: **Do not stop the forward motion or cause the tracks to spin while pushing material.**

Step 4. Make successive cuts the same as in step 2, constructing the stockpile higher on each pass until it reaches the desired height.

SPREADING A STOCKPILE

2-31. Large piles should be worked from the side, cutting material away from the stockpile, using one-third of the blade. Use the following steps to spread a stockpile:

Step 1. Lower the blade to the desired height while moving forward.

Step 2. Adjust the blade height and move the dozer into the side of the pile making the cut with only one-third of the blade.

NOTE: **When using the left side of the blade, continue working to the left. When using the right side of the blade, continue working to the right.**

Step 3. Cut into the stockpile. The blade should be as full as possible without stalling the dozer or spinning the tracks. Raise and lower the blade to maintain a smooth pass.

WARNING

When spreading materials that are higher than the midpoint of the rollover protective structure (ROPS), adjust the height of the cut to eliminate the danger from collapsing material.

Step 4. Spread the blade load after cutting the pile by continuing to move forward and slowly raising the blade until all material is evenly feathered.

Step 5. Feather the blade load and reverse the dozer. Raise the blade about 12 inches off the ground, back the dozer to the stockpile, and reposition for another cut.

Repeat the above steps until the stockpile has been leveled and spread over the designated area. **Do not** back blade to level the stockpile.

BACKFILLING

2-32. Backfilling can be effectively accomplished by drifting material sideways with an angle blade. This allows forward motion parallel to the excavation. With a straight blade, approach the excavation at a slight angle and then, at the end of the pass, turn in toward the excavation. No part of the tracks should hang over the edge. Adjust the length of the push based on soil conditions. For example, when working in soft material or on an unstable slope, let the second bladeful push the first bladeful over the edge. Be careful to keep oversize materials out of the backfill.

RIPPING

2-33. *Figure 2-12* shows various ripping operations. Use first gear for ripping operations. When performing one-shank ripping, always use the center shank. Use additional shanks, where practical, to increase production. When ripping for scraper loading, rip in the same direction that the scrapers are loading, whenever possible. It is usually desirable to rip as deeply as possible. However, it is sometimes better to rip the material in its natural layers even if this is less than full-shank depth. Use the ripped material on top of the unripped formation to cushion the machine and provide traction. When the final material size must be relatively small, space passes close together. Cross rip only when necessary to obtain the required breakage. Use the following steps to rip material:

Step 1. Position the dozer on the uphill side if operating on a slope, about half the length of the dozer from the start of the area to be ripped.

Step 2. Place the transmission shift lever in forward, first gear.

Step 3. Lower the rippers to the ripping depth as the dozer begins to cross the area to be ripped.

WARNING

Maintain a straight line while ripping. Turning the dozer with the rippers in the ground will cause damage to the dozer.

Step 4. Raise the rippers out of the ground and then stop at the end of the pass.

Step 5. Place the transmission in reverse and back the dozer to the start point.

Step 6. Position the dozer to overlap the previous ripping pass.

Repeat steps 1 through 6 until the area is completely ripped.

Packed Soil, Hardpan, Shale, and Cemented Gravel

2-34. Three-shank ripping works well in these materials. Use as many shanks as possible to break material to the desired size.

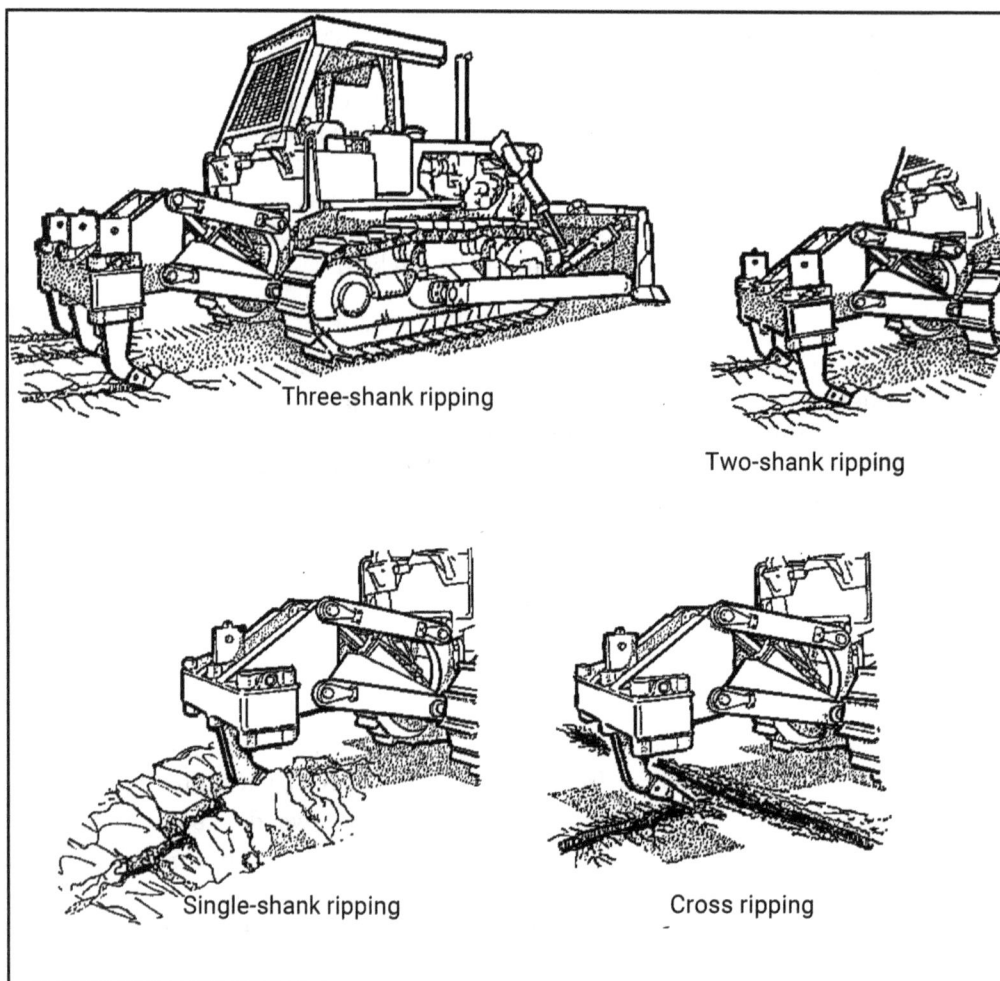

Figure 2-12. Ripping Operations

Rock with Fractures, Faults, and Planes of Weakness

2-35. Use two shanks for ripping where rocks break out in small pieces and the machine can handle the job easily. Use only the center shank if the machine begins to stall or the tracks spin.

- **Asphalt.** Raise the ripper shank to lift out and break the material.
- **Concrete.** Use one-shank ripping to sever reinforcing rods or wire mesh effectively.

Solid Rock, Granite, and Hard-to-Rip Material

2-36. Use one shank in hard-to-rip material or material that tends to break out in large slabs or pieces.

WINCHING

2-37. Winching is hoisting or hauling with a winch, using a cable. When winching, make sure personnel are clear of the cable. Cables can break and cause severe injury. Exercise caution with suspended loads. If the engine revolutions (speed) are too low, the weight of the load may exceed the engine capacity causing the load to drop, even though the winch is in the reel-in position.

> **CAUTION**
> Always keep the winch cable in a straight line behind the machine. For safety and maximum service life of the winch component, decelerate the engine before moving the winch control lever. After shifting, control the cable speed by varying the engine speed. Winch loads at low engine speed with the machine stationary. When moving away from a load, operate the machine in low gear to prevent overspeeding of winch components. Do not operate the winch for extended durations.

DOZER PRODUCTION ESTIMATES

2-38. Dozer production curves give maximum-production values (in LCY per hour) for straight and universal blades based on the following conditions:

- A 60-minute working hour (100 percent efficiency).
- Power-shift machines with 0.05-minute fixed times are being used.
- The dozer cuts 50 feet, then drifts the blade load to dump over a high wall.
- The soil density is 2,300 pounds per LCY.
- The coefficient of traction equals 0.5 or better for crawler machines and 0.4 or better for wheel machines.
- Hydraulic-controlled blades are being used.

2-39. Use the following steps to estimate dozer production:

Step 1. Determine the maximum production. Determine the estimated maximum production from either *Figure 2-13* or *2-14*, based on the type of dozer being used. Find the dozing distance on the bottom horizontal scale in the proper figure. Read up vertically until intersecting the production curve for the dozer being considered then read the vertical scale on the left to determine the maximum production in LCY per hour.

- Use *Figure 2-13* to determine the estimated maximum production for D3 through D6 tractors with straight blades. The DEUCE has the same production capability as the D5.
- Use *Figure 2-14* to determine the estimated maximum production for D7 or D8 tractors with universal or straight blades.

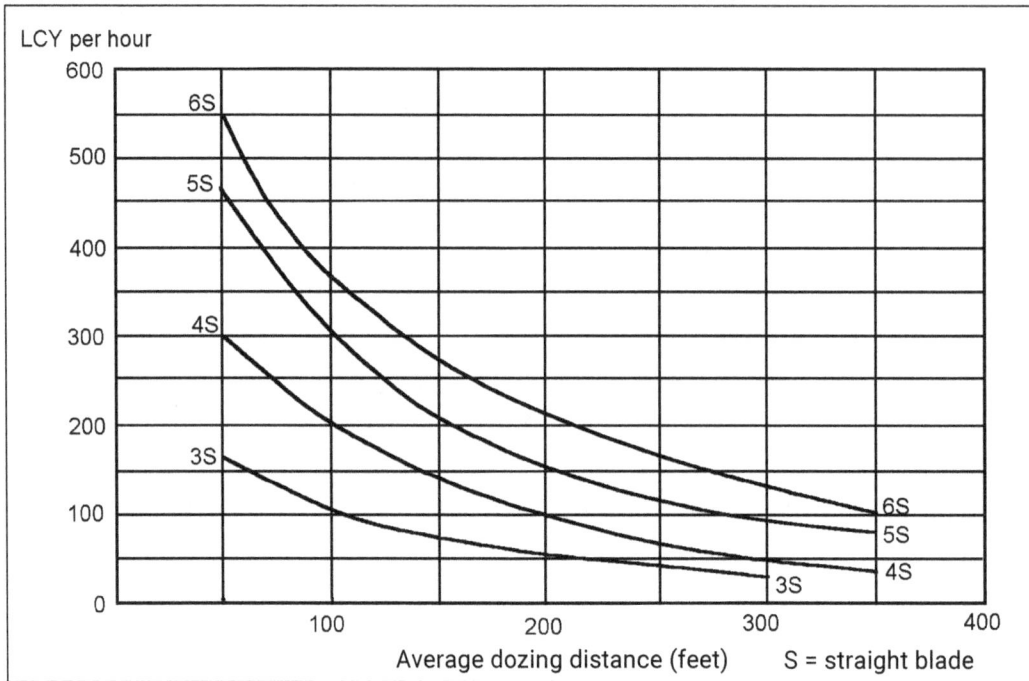

Figure 2-13. Estimated Maximum Production for D3 Through D6 Tractors With Straight Blades

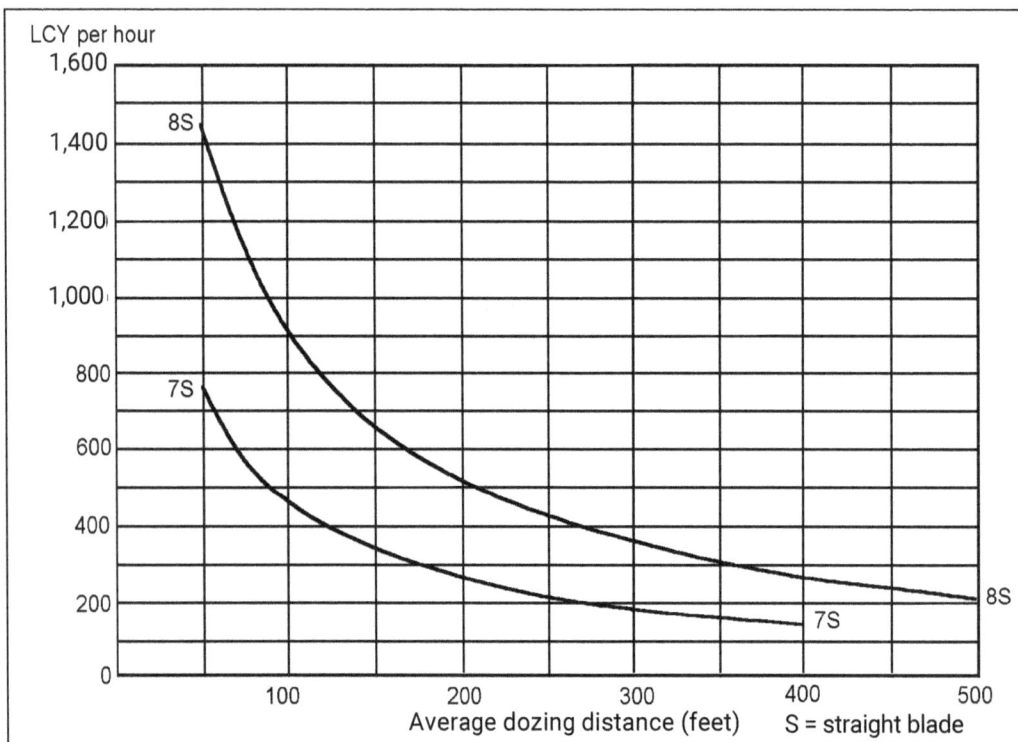

Figure 2-14. Estimated Maximum Production for D7 or D8 Tractors With Universal or Straight Blades

Step 2. Determine the grade correction factor—(-) favorable or (+) unfavorable. Find the percent grade on the top horizontal scale of *Figure 2-15*. Read down vertically and intersect the grade correction curve, then read to the right horizontally and locate the grade correction factor on the vertical scale.

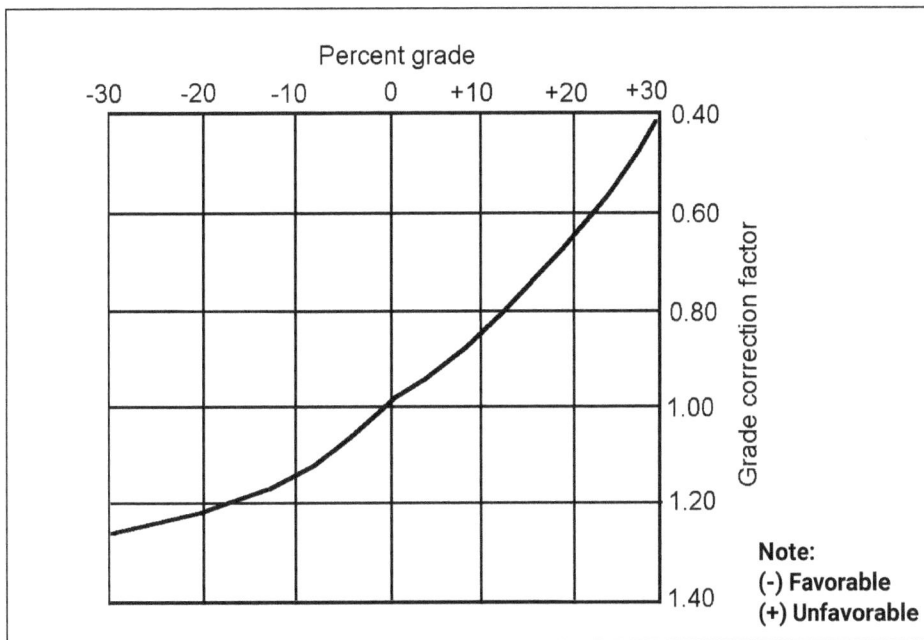

Figure 2-15. Dozer-Production Grade Correction Factors

Step 3. Determine the material-weight correction factor. If the actual unit weight of the material to be pushed is not available from soil investigations, use the average values found in *Table 1-2, page 1-4*. Divide 2,300 pounds per LCY by the material's LCY weight to find the correction factor. Soil density of 2,300 pounds per LCY is a constant that was assumed in determining the maximum production.

$$\text{Material-weight correction factor} = \frac{\text{2,300 pounds per LCY (standard material unit weight)}}{\text{actual material LCY weight}}$$

where—
2,300 = standard material unit weight per LCY

Step 4. Determine the material-type correction factor. Dozer blades are designed to cut material and give it a rolling effect in front of the blade. This results in a production factor of 1. *Table 2-4* gives the correction factors to account for how different materials behave in front of the blade.

Table 2-4. Material-Type Correction Factors

Material State	Factor for Crawler Tractors
Loose, stockpile	1.2
Hard to cut; frozen, with tilt cylinder Hard to cut; frozen, without tilt cylinder	0.8 0.7
Hard to drift; dead (dry, noncohesive) material or very sticky material	0.8
Rock (ripped or blasted)	0.6 to 0.8

Step 5. Determine the operator correction factor (see *Table 2-2, page 2-7*).

Step 6. Determine the operating-technique correction factor from *Table 2-5*.

Table 2-5. Operating-Technique Correction Factors

Operating Technique	Factor for Crawler Tractors
Slot dozing	1.2
Side-by-side dozing	1.15 to 1.25

Step 7. Determine the efficiency factor. In the case of a well-managed job, expect 50 minutes of production time per hour.

$$\text{Efficiency factor} = \frac{\text{actual working minutes per hour}}{\text{60-minute working hour}}$$

Step 8. Determine dozer production.

Production (LCY per hour) = maximum production the product of the correction factors

Step 9. Determine the material conversion factor, if required. To find the total time (step 10) and the total number of dozers required to complete a mission within a given time (step 11), adjust the volume of material that is being moved and the equipment production rate per hour so that they both represent the same material state. Refer to material and production states as LCY, BCY, and CCY. If necessary to convert, use *Table 1-1, page 1-4,* to find the material conversion factor. Multiply the conversion factor by the production per hour to find the production per hour in a different state.

NOTE: This conversion will not change the dozer production effort.

EXAMPLE

Determine the average hourly production (in CCY) of a straight-blade D7 (with tilt cylinder) moving hard-packed clay an average distance of 200 feet, down a 10 percent grade, using slot dozing. Estimated material weight is 2,500 pounds per LCY. The operator is of average ability and will work during daylight hours. Expected efficiency is 50 minutes per hour.

Step 1. Uncorrected maximum production = 300 LCY per hour   *(Figure 2-14, page 2-19)*

Step 2. Grade correction factor = 1.15   *(Figure 2-15, page 2-20)*

Step 3. Material-weight correction factor

$$= \frac{2{,}300 \text{ pounds per LCY (standard material unit weight)}}{2{,}500 \text{ pounds per LCY (actual material unit weight)}}$$

$$= 0.92$$

Step 4. Material-type correction factor (a hard-to-cut material) = 0.8   *(Table 2-4, page 2-21)*

Step 5. Operator correction factor = 0.75   *(Table 2-2, page 2-7)*

Step 6. Operating-technique correction factor = 1.2   *(Table 2-5, page 2-21)*

Step 7. Efficiency factor $= \dfrac{50 \text{ working minutes per hour}}{60\text{-minute working hour}} = 0.83$

Step 8. Dozer production

$$= 300 \text{ LCY per hour} \times 1.15 \times 0.92 \times 0.8 \times 0.75 \times 1.2 \times 0.83$$
$$= 190 \text{ LCY per hour per dozer}$$

Step 9. Material conversion factor = 0.63

Dozer production in CCY $= 0.63 \times 190 \text{ LCY per hour} = 120 \text{ CCY per hour}$

Step 10. Determine the total time required in hours.

$$\text{Total time (hours)} = \frac{Q}{P \times N}$$

where—
Q = quantity of material to be moved
P = hourly production rate per dozer
N = number of dozers

EXAMPLE

Determine the total time required to move 3,000 CCY of hard-packed clay, using one D7 dozer with a production rate of 120 CCY per hour.

$$\frac{3{,}000 \text{ CCY}}{120 \text{ CCY per hour} \times 1 \text{ dozen}} = 25 \text{ hours}$$

Step 11. Determine the total number of dozers required to complete the mission within a given time.

$$\text{Total number of dozers} = \frac{Q}{P \times T}$$

where—
    Q = quantity of material to be moved
    P = hourly production rate per dozer
    T = maximum allowable duration, in hours

EXAMPLE

Determine how many D7 dozers (with a production rate of 120 CCY per hour) would be needed to move 3,000 CCY of clay in seven hours.

$$\frac{3,000 \text{ CCY}}{120 \text{ CCY per hour} \times 7 \text{ hours}} = 3.6 \text{ D7 dozers (round up to 4 dozers)}$$

RIPPING PRODUCTION ESTIMATES

2-40. The best method to estimate ripping production is by working a test section and recording the time required and the production achieved. However, the opportunity to conduct such investigations is often nonexistent and, therefore, estimates are usually based on historical production charts. Ripping applications will increase the machine's maintenance requirements by 30 to 40 percent.

QUICK METHOD

2-41. A quick method to determine an approximate production rate is to time several passes of a ripper over a measured distance. The timed duration should include the turnaround time at the end of the pass. Determine an average cycle time from the timed cycles. Determine the quantity (volume) from the measured length multiplied by the width of the ripped area and the depth of penetration. If measurements are in feet, divide the number of feet by 27 to convert cubic feet to cubic yards.

$$\text{Volume BCY} = \frac{\text{length (feet)} \times \text{width (feet)} \times \text{penetration depth (feet)}}{27}$$

where—
    27 = factor used to convert cubic feet to cubic yards

2-42. Experience has shown that the production rate calculated by this quick method is about 20 percent higher than an accurately cross-sectioned study. Therefore, the formula for estimating ripping production is—

$$\text{Ripping production (BCY per hour)} = \frac{V}{T \times 1.2}$$

where—
    V   = measured volume in BCY
    T   = average time in hours
    1.2 = method correction factor

SEISMIC-VELOCITY METHOD

2-43. Most ripping-production charts are based on the relationship between the ripability and the seismic-wave velocity response of a material. The *Figure 2-16* ripping performance chart, which is for a 300-horsepower dozer, allows the estimator to make a determination of the machine's performance capability based on seismic velocity and general rock classifications. After establishing a seismic velocity, estimate production from the production chart in *Figure 2-17*. This chart provides a band of production rates representing ideal-to-adverse rock conditions based on the following assumptions:

- The efficiency factor is 100 percent (60-minute working hour).
- The power-shift machines used have single-shank rippers.
- The machine rips full-time, no dozing.
- The upper limit of the band reflects ripping under ideal conditions only. If conditions such as thick laminations, vertical laminations, or other rock structural conditions exist which would adversely affect production, use the lower limit.

2-44. Regardless of the seismic velocity, tooth penetration is the key to ripping success. This is particularly true for homogeneous materials such as mudstone, clay stone, and fine-grained caliches.

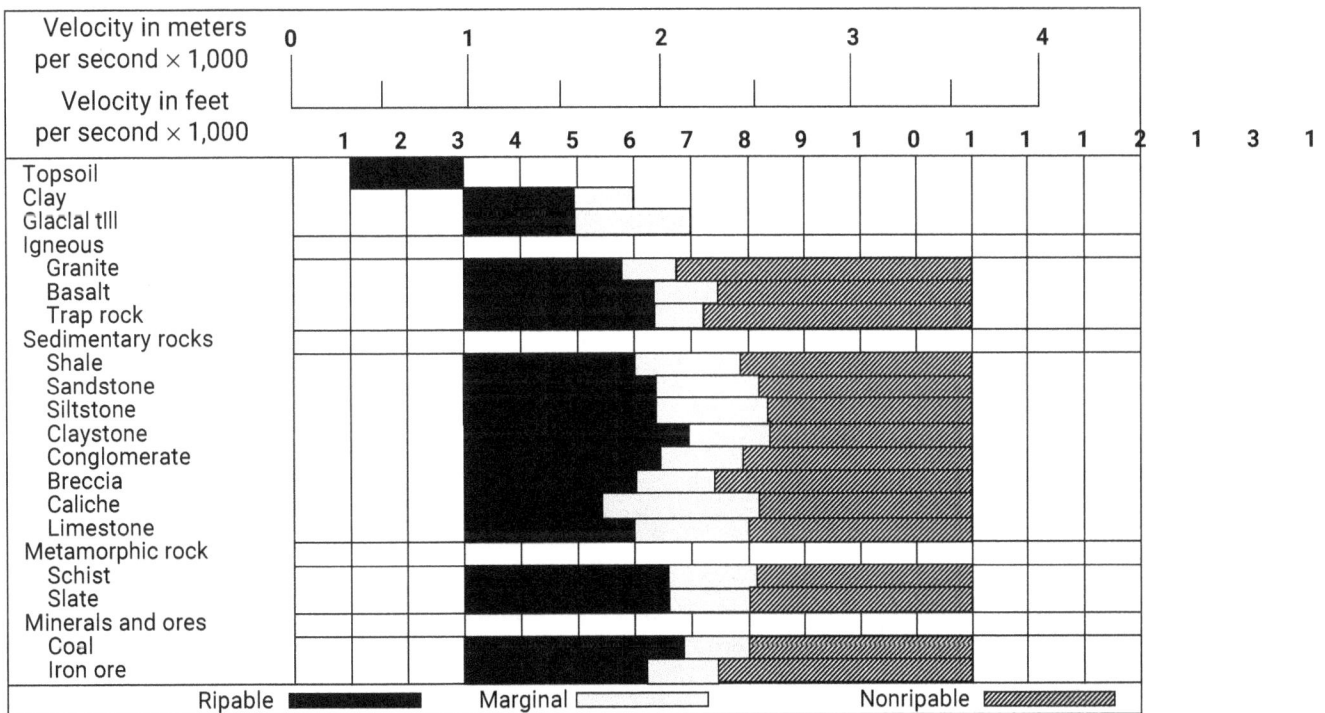

Figure 2-16. Ripping Performance for a 300-Horsepower Dozer
With a Single- or Multishank Ripper

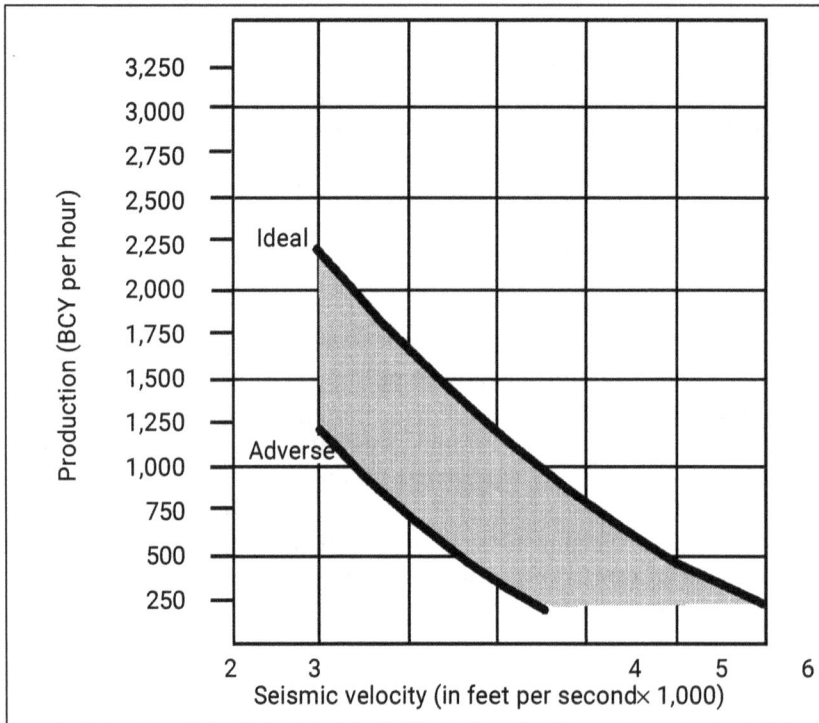

Figure 2-17. Estimated Ripping Production for a 300-Horsepower Dozer With a Single-Shank Ripper

Production (BCY per hour per dozer)   P E

where—
P = maximum production for a 300-horsepower dozer (*Figure 2-17*)
E = efficiency factor

NOTE: Before referring to *Figure 2-17* for determining a probable production rate, refer to *Figure 2-16* to verify the ripability with the equipment available.

EXAMPLE

Determine how many 300-horsepower dozers are needed to rip 9,000 BCY of limestone having a seismic velocity of 4,000 feet per second in 7 hours. The limestone is bedded in thin laminated layers. Efficiency will be a 45-minute working hour.

Maximum production for ideal conditions (thin layers) is 1,700 BCY per hour *(Figure 2-17).*

Efficiency-adjusted production

$$= 1,700 \text{ BCY per hour} \times \frac{45}{60}$$

$$= 1,275 \text{ BCY per hour}$$

$$\frac{9,000 \text{ BCY}}{1,275 \text{ BCY per hour} \times 7 \text{ hours}} = 1,300\text{-horsepower dozer}$$

SAFETY PRECAUTIONS

2-45. Listed below are some specific safety precautions for dozer operators:

- Never carry personnel on the tractor drawbar.
- Never turn around on steep slopes; back up or down instead.
- Keep the machine in low gear when towing a heavy load downhill.
- Always lower the blades when the machine is parked.
- Ensure that only one person is on the machine while it is in operation. However, in some training situations it is necessary to have two people on a dozer while it is in operation.

Chapter 3

Scrapers

The design of scrapers (tractor scrapers) allows for loading, hauling, dumping, and spreading of loose materials. Use a scraper for medium-haul earthmoving operations and for moving ripped materials and shot rock. The haul distance (zone of operation), the load volume, and the type and grade of surface traveled on are the primary factors in determining whether to use a scraper on a particular job. The optimum haul distance for small- and medium-size scrapers is 3,000 feet or less.

DESCRIPTION

3-46. *Figure 3-1, page 3-2,* shows a CAT® 621B single-powered-axle wheel scraper. The CAT 621 is designed to operate using a push tractor for loading assistance. The air-droppable CAT 613B wheel scraper has a chain-elevator loading mechanism that allows it to load without the assistance of a push tractor. The basic operating parts of a scraper are these:

- **Bowl.** The bowl is the loading and carrying component. It has a cutting edge, which extends across the front bottom edge. Lower the bowl until the cutting edge enters the ground for loading, raise it for carrying, and lower it to the desired lift thickness for dumping and spreading.
- **Apron.** The apron is the front wall of the bowl. It is independent of the bowl and, when raised, it provides an opening for loading and spreading. Lower the apron during hauling to prevent spillage.
- **Ejector.** The ejector is the rear wall of the bowl. Keep the ejector in the rear position when loading and hauling materials. Activate the ejector to move forward during spreading to provide positive discharge of materials.

CAPACITY

3-47. *Struck* capacity means the bowl has a full load of material that is level with its sides. *Heaped* capacity means the material is heaped in the bowl and slopes down on a 1:1 repose slope to the sides of the bowl. In practice, these will be LCY of material because of how a scraper loads. Therefore, load volume in terms of BCY moved depends on both the bowl size and the material type being loaded. The rated volumetric capacity of the Army 621B scraper is 14-cubic-yards struck and 20-cubic-yards heaped. The capacity of the CAT 613B scraper is 11-cubic-yards heaped. Elevating scrapers, like the Army 613, are not given struck capacity ratings.

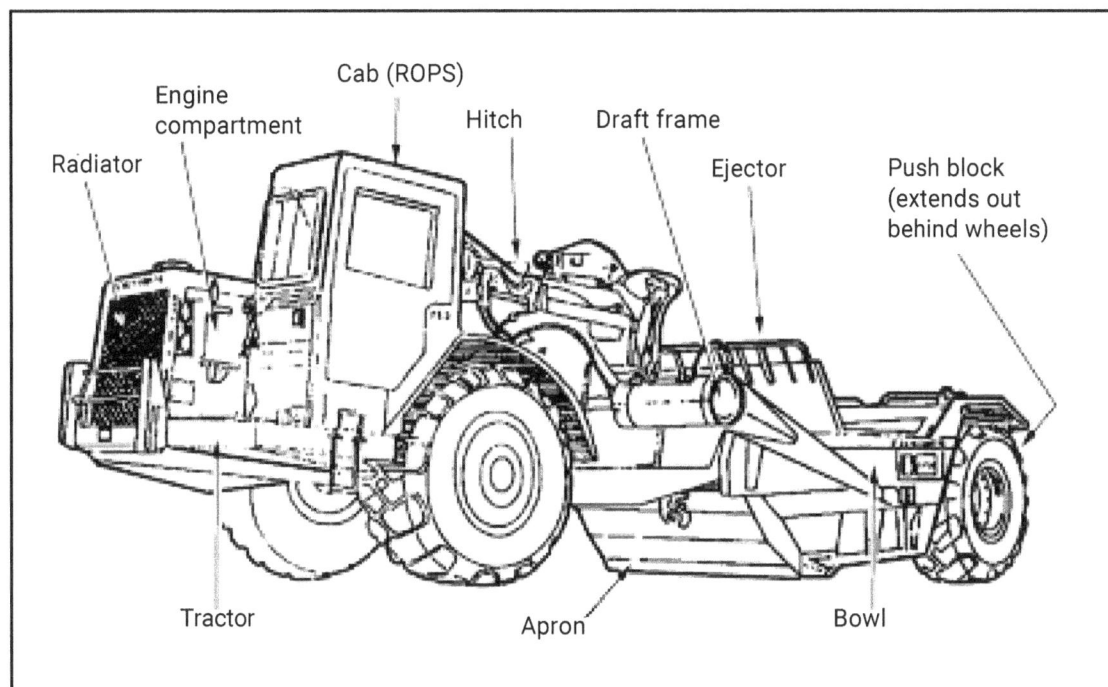

Figure 3-18. CAT 621B Wheel Scraper

OPERATING RANGE

3-48. The optimum haul distance for the small- and medium-size scrapers is 300 to 3,000 feet. There are larger scrapers that are effective up to 5,000 feet.

SELECTION

3-49. A scraper is a compromise between a machine designed exclusively for either loading or hauling. For medium-distance movement of material, a scraper is better than a dozer because of its travel-speed advantage and it is better than a truck because of its fast load time, typically less than a minute. Another advantage of the scraper is that it can spread its own load and quickly complete the dump cycle.

PRODUCTION CYCLE

3-50. The production cycle for a scraper consists of six operations—loading, haul travel, dumping and spreading, turning at the dump site, return travel, and turning and positioning to load. *Figure 3-2* shows the functions of the apron, bowl, and ejector during loading, hauling, and dumping.

LOADING

3-51. The CAT 621 loads with push-tractor assistance. This scraper can load to a limited extent without assistance, but requires push loading to achieve maximum production. Pusher assistance is necessary to reduce loading time and wheel spinning. Reducing scraper wheel spinning increases tire life. The scraper should not depend on the pusher to do all the work. Conversely, do not spin the scraper's wheels to pull away from the pusher. Use pusher assistance for either straight, downhill, or straddle loading. Always load the scraper in the direction of haul. Do not turn the scraper at the same time it is accelerating from the loading operation. The CAT 613 is a self-loading machine, and pushing during loading will damage the scraper's loading elevator.

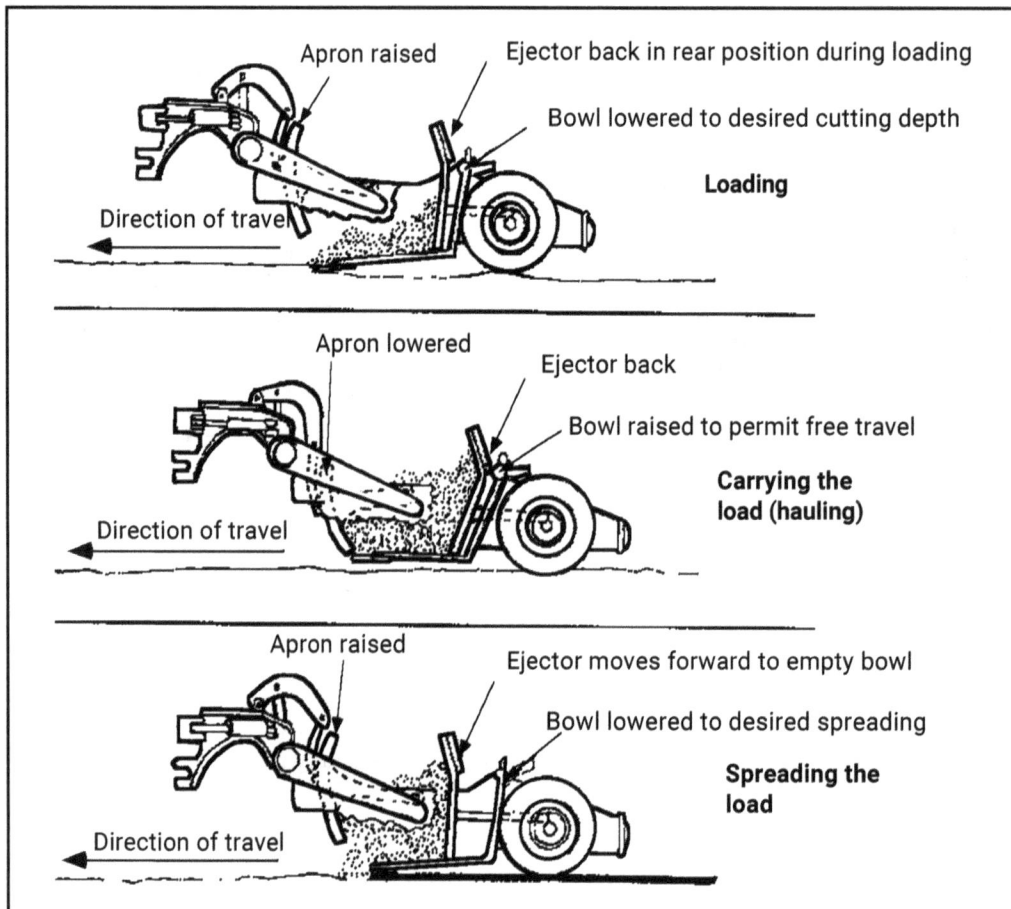

Figure 3-19. Functions of the Apron, Bowl, and Ejector

Downhill Loading

3-52. Downhill loading enables a scraper to obtain larger loads in less time. Each 1 percent of favorable grade is equivalent to increasing the loading force by 20 pounds per ton of gross scraper weight.

Straddle Loading

3-53. Straddle loading (*Figure 3-3, page 3-4*) requires three cuts with a scraper. The first two cuts should be parallel, leaving a ridge between the two cuts. The scraper straddles this ridge of earth to make the final cut. The ridge should be no wider than the distance between a scraper's wheels. With straddle loading, time is gained on every third trip because the center strip loads with less resistance than a full cut.

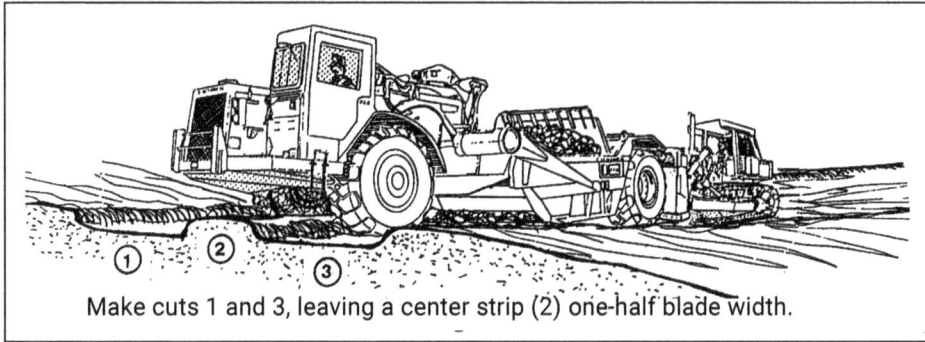

Make cuts 1 and 3, leaving a center strip (2) one-half blade width.

Figure 3-20. Straddle Loading With Pusher Assistance

Push-Loading

3-54. **Back-Track.** Use the back-track push-loading technique *(Figure 3-4)* where it is impractical to load in both directions. However, this method is inefficient due to the time spent in backing up and repositioning for the next load.

3-55. **Chain.** Use the chain push-loading technique *(Figure 3-4)* where the cut is fairly long, making it possible to pick up two or more scraper loads without backtracking. The pusher pushes one scraper, then moves behind another scraper that is moving in the same direction in an adjacent lane.

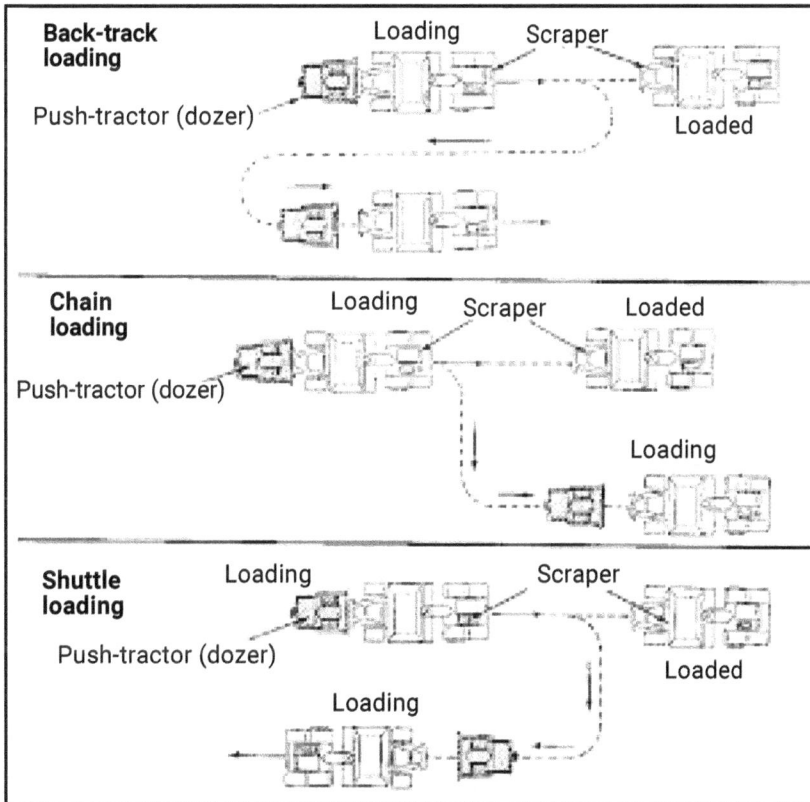

Figure 3-21. Push-Loading Techniques

3-56. **Shuttle.** Use the shuttle push-loading technique *(Figure 3-4)* for short cuts where it is possible to load in both directions. The pusher pushes one scraper, then turns and pushes a second scraper in the opposite direction.

Cut-and-Load Sequence

3-57. The scraper loading sequence is as follows:

Step 1. Use the service brake to reduce scraper travel speed when close to the cut (loading lane), and downshift to first gear for loading.

Step 2. Move the ejector to the rear.

Step 3. Open the apron partway.

Step 4. Lower the bowl to an efficient cut depth after the scraper enters the cut. Continue moving forward until the dozer contacts the scraper and begins pushing. If the scraper tires spin before the dozer makes contact, stop and allow the dozer to assist. When the dozer makes contact, push down both the differential lock and the transmission hold pedal and proceed in second gear. The cut should be as deep as possible, but it should allow the scraper to move forward at a constant speed without lugging the engine. Decrease the cut depth if the scraper or pusher lugs or if the drive wheels slip. Use the router bits on the vertical side of the bowl to gauge the depth of cut. Once an efficient depth of cut is determined, use that same depth on successive passes.

Step 5. Mark the cut. When cutting—

- Regulate the apron opening to prevent material from piling up in front of the lip or falling out of the bowl.
- Keep the machine moving in a straight line while maintaining pusher and scraper alignment.
- Do not overload the scraper. Overloading lowers efficiency and places unnecessary stresses on the machine.
- Raise and lower the bowl rapidly when loading loose material such as sand.

NOTE: When a push tractor is used, it should be waiting about 45 ° off of the lane to be cut. This allows the loading unit to come in with the least delay and difficulty.

Step 6. Raise the bowl slowly when full, while at the same time closing the apron to prevent spillage.

Step 7. Allow the pusher to help the machine out of the cut area, if necessary.

NOTE: When exiting the cut, release the transmission hold and/or the differential lock, if in use. Accelerate to travel speed as quickly as possible. Travel a few feet before lifting the bowl to the carrying position. This spreads any loose material piled up in front of the bowl and allows the following scraper to maintain speed.

Materials

3-58. **Loam and Clay.** . Loam and most clay soils cut easily and rapidly with minimum effort. However, loosen very hard clay with a dozer ripper before loading.

3-59. **Sand.** Since sand has little or no cohesion between its particles, it has a tendency to run ahead of the scraper blade and apron. The condition is worse for finer and drier particles. When loading sand, the best method is as follows:

Step 1. Enter the loading area fast, lowering the bowl slowly, and pick up as much material as possible using the momentum of the scraper unit. This will fill the hard-to-reach rear area of the bowl.

Step 2. Shift to a lower gear once the momentum is lost, and allow the pusher to assist.

Step 3. Pump the bowl up and down *(Figure 3-5).* For best pumping results, drop the bowl as the scraper's rear wheels roll into the depression of the previously pumped area and raise the bowl as the wheels are climbing out of the depression.

Step 4. Drop the bowl sharply two or three times at the end of the loading area to top out the load. Then close the apron, raise the bowl, and exit the cut area.

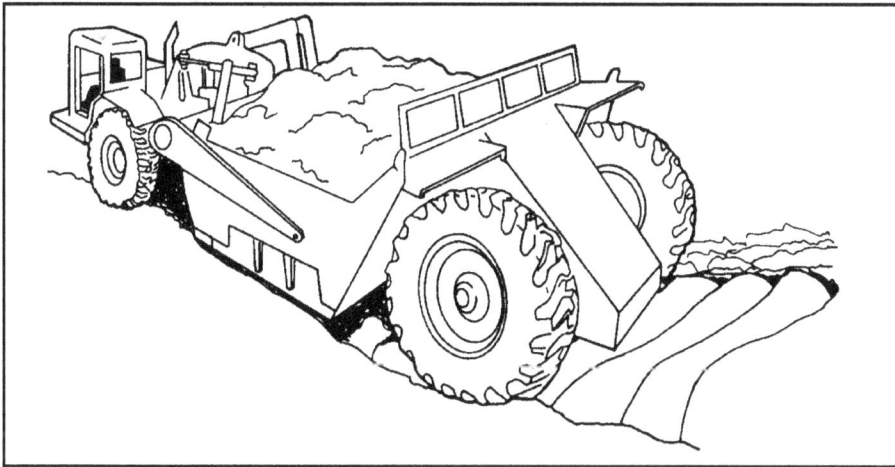

Figure 3-22. Pumping a Scraper Bowl to Load Sand

3-60. **Rock and Shale.** Loading rock and shale with a scraper is a difficult task that causes severe wear and tear on the equipment. Ripping will ease this problem. Ripping depth should exceed the depth of the scraper cut. When loading the scraper, leave a loose layer of ripped material under the tires to provide better traction and to reduce both track and tire wear. Some soft rock and shale can be loaded without ripping.

3-61. Start the scraper's cutting edge in dirt (if possible) when loading stratified rock. Move in to catch the blade in planes of lamination. This forces material into the bowl. Pick up loose rock or shale on the level or on a slight upgrade, with the blade following the lamination planes.

Load Time

3-62. Loading time is critical for obtaining maximum scraper production. Push loading should normally take less than one minute within a distance of 100 feet (time and distance change with the material being loaded). Studies of load volume versus loading time indicate that for a normal operation, about 85 percent of scraper load capacity is achieved in the first 0.5 minute of loading. Another 0.5 minute will only produce about another 12 percent increase in load volume. Therefore, extra loading time (past about one minute) is not worth the effect because increased total cycle time will decrease production.

Borrow-Pit Operation

3-63. It is essential to have highly competent personnel in the borrow-pit area. Traffic control within the borrow-pit area reduces waiting time and excess travel of earthmoving support units. Maintaining adequate drainage throughout the borrow pit will reduce downtime caused by bad weather.

HAULING

3-64. Hauling, or travel time, includes the haul time and the return time. Here the power and traction characteristics of the scraper become very important. The following factors can greatly effect travel time.

Haul-Route Location

3-65. Lay out the haul routes to eliminate unnecessary maneuvering. Plan the job to avoid adverse grades that could drastically reduce production. Remember, where grades permit, the shortest distance between two points is always a straight line.

Road Maintenance

3-66. Keep haul roads in good condition. A well-maintained haul road permits traveling at higher speeds, increases safety, and reduces operator fatigue and equipment wear.

- **Ruts and rough surfaces.** Use a grader or dozer to eliminate ruts and rough (washboard) surfaces. (See *Chapter 4* for haul-road maintenance with a grader.)
- **Dust.** Use water distributors to reduce dust. Reducing the amount of dust helps alleviate additional mechanical wear, provides better visibility, and lessens the chance of accidents. Keeping roads moist (but not wet) allows them to pack into hard, smooth surfaces permitting higher travel speeds.

Travel Conditions

3-67. Once on the haul road, the scraper should travel in the highest safe gear appropriate for road conditions. When possible, carry the scraper bowl fairly close to the ground (about 18 inches). This lowers the center of gravity of the scraper and reduces the chance of overturning.

- **Lugging.** Avoid unnecessary lugging of the engine. Downshift when losing momentum. Lugging the engine usually results in a slower speed than the top range of the next lower gear. Although the machine can make it, it is best to downshift and accelerate faster. Lugging causes a decrease in engine revolutions per minute (rpm) thereby reducing hydraulic pressure. This will result in a loss of steering control.
- **Coasting.** Never coast on a downgrade. When approaching a downgrade, slow down and downshift the transmission. To prevent unwanted upshifting, use the transmission hold on a downgrade if it is available. Also, use it when approaching an upgrade or in rough

underfooting. To control speed on a downgrade, use the retarder and service brakes. Engine speed should not exceed the manufacturer's recommended rpm.

DUMPING AND SPREADING

3-68. When dumping, lower the bowl to the desired lift height and open the apron at the beginning of the dump area. Dump and spread in the highest gear permitted by haul-road conditions and fill-material characteristics. Constant speed, along with bowl height, will help to maintain a uniform depth of lift. Slowly dribbling the load at low speed slows down the cycle.

Dumping Procedure

Step 1. Move steadily across the spreading area.

Step 2. Open the apron fully as the scraper reaches the location to begin dumping. Move the ejector forward to push the material out of the bowl.

Step 3. Maintain a straight path through the spread area.

Step 4. Close the apron when all the material is out of the bowl, and return the ejector to the rear of the bowl.

Step 5. Raise the bowl slowly to clear obstacles (12 to 18 inches) during the return trip to the loading area.

Spreading Sequence *(Figure 3-6)*

Step 1. Dump and spread the first load at the front of the fill.

Step 2. Travel with subsequent loads over the previous fill, provided the lifts are shallow.

Step 3. Start each following dump at the end of the previous fill.

Step 4. Finish dumping and spreading one full lane before starting a new one so that rollers can start compaction.

Step 5. Repeat this method in the next lane. Do not waste time on the fill. The scraper should return to the cut area as fast as possible after dumping the load. Plan the egress from the fill area to avoid soft ground or detours around trees or other obstacles.

NOTE: Route the scrapers to compact the fill. Overlap wheel paths to aid in compaction of the entire area and to reduce compaction time for rollers.

CAUTION
Do not try to force wet or sticky material out of the bowl too fast. This will cause the material to roll up in front, which can damage the hydraulic system.

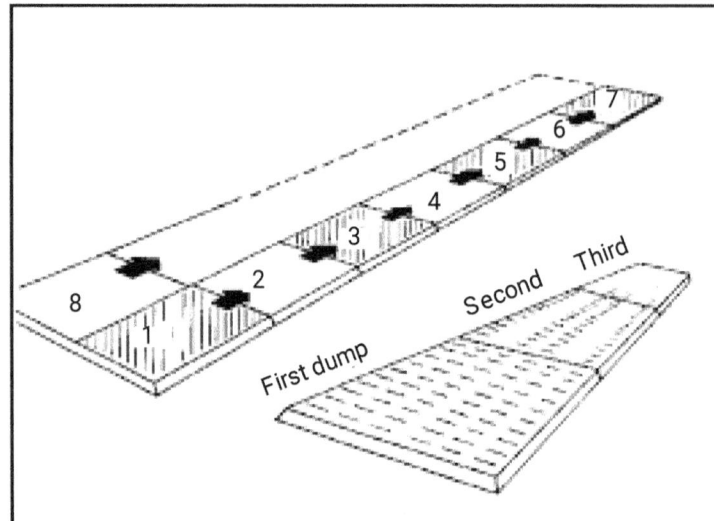

Figure 3-23. Spreading Sequence

Fill Slope

3-69. To maintain the desired fill slope, make the fill high on the outside edges. This will prevent the scraper from sliding over the slope and damaging the slope. If there is rain, build up the low center for drainage, or use a grader to cut the outside edge down, creating a crown in the middle of the area.

Materials

3-70. Different materials require different dumping and spreading procedures.

3-71. **Sand.** Spread sand as thin as possible to allow better compaction and to make traveling over the fill easier.

3-72. **Wet or Sticky Material.** Wet or sticky material may be difficult to unload or spread. When operating in these material types—

- Do not try to spread the material too thin.
- Keep the bowl high enough to allow the material to pass under the scraper. Material not having enough room to pass under the scraper will roll up inside the bowl into a solid mass that is difficult to eject.
- Bring the ejector forward about 12 inches at a time.
- Back the ejector about 6 inches after each forward movement. This breaks the suction between the material and the bowl.
- Repeat this procedure until the bowl is empty.

PRODUCTION ESTIMATES

3-73. Following is an explanation of production estimating based on a step-by-step method using the CAT 621B scraper. When developing data for production estimates, consider all factors that influence production. Consider the scraper's weight, the weight of the load, and the average grade and rolling resistance of both the haul and return routes in arriving at a cycle time. The same steps are applicable to other makes and models of scrapers, using the appropriate tables and charts for those scrapers.

Step 1. Determine the vehicle weight, empty and loaded.

- **Empty vehicle weight (EVW), in tons.** Using *Table 3-1,* first determine the EVW from the EVW column based on the specific make and model of the scraper.

- **Weight of load, in tons.** Determine the weight of the load in pounds by determining the scraper load volume in cubic yards (this is in LCY of the material) and the material unit weight (in pounds per LCY). If no specific material-weight data is available, use the information in *Table 1-2, page 1-4,* as an estimate. Multiply the scraper load volume by the unit weight in pounds per LCY of the material to be excavated. Then, convert the resulting weight into tons by dividing the amount by 2,000.

$$\text{Weight of load (pounds)} = \text{scraper load volume (LCY)} \times \text{material unit weight (pounds per LCY)}$$

$$\text{Weight of load (tons)} = \frac{\text{weight of load (pounds)}}{2,000}$$

Table 3-6. Scraper Specifications

Make and Model	Heaped Capacity (Cubic Yards)	EVW (Tons)
CAT 613B	11	15.6
CAT 621B	20	33.3

- **Loaded or gross vehicle weight (GVW).** Determine the GVW by adding the EVW (tons) and the weight of load (tons).

$$\text{GVW (tons)} = \text{EVW (tons)} + \text{weight of load (tons)}$$

EXAMPLE

Determine the GVW of a CAT 621B single-powered scraper with a 20 LCY load of dry loam.

From *Table 1-2,* dry loam is 1,900 to 2,200 pounds per LCY. Use an average value of 2,050 pounds per LCY.

$$\text{Weight of load (tons)} = \frac{20 \text{ LCY} \times 2,050 \text{ pounds per LCY}}{2,000 \text{ pounds per ton}} = 20.5 \text{ tons}$$

$$\text{EVW} = 33.3 \text{ tons}$$

$$\text{GVW} = 33.3 \text{ tons} + 20.5 \text{ tons} = 53.8 \text{ tons}$$

Step 2. Determine the average grade (in percent) and the distance (in feet) for both the haul and return routes. Uphill grades are positive (+) and downhill grades are negative (-). Obtain this information from a mass diagram or a haul-route profile.

EXAMPLE

The project mass diagram indicates that there is a 5 percent downhill grade from cut to fill and that the one-way distance is 800 feet. The same route will be used for both the haul and the return.

Haul:
   Average grade = -5 percent
   Distance = 800 feet
Return:
   Average grade = +5 percent
   Distance = 800 feet

Step 3. Determine the rolling resistance (in pounds). Rolling resistance is the force resisting the movement of a vehicle on level ground. This is primarily caused by the tires penetrating the road's surface, the tires flexing, and internal gear friction *(Figure3-7, page 3-12)* . Express the rolling resistance for a given road surface in pounds per ton of vehicle weight.   *Table 3-2, page 3-12,* gives some representative rolling-resistance values for various types of road surfaces. If the expected tire penetration is known, determine the rolling resistance for the haul and the return using the following formulas:

$$RR_{Haul} = (40 + [30 \times TP]) \times GVW$$

$$RR_{Return} = (40 + [30 \times TP]) \times EVW$$

where—

RR_{Haul} = haul rolling resistance, in pounds

RR_{Return} = return rolling resistance, in pounds

40 = constant that represents the flexing of the driving mechanism, in pounds per ton

30 = constant that represents the force required to climb out of the rut, in pounds per ton per inch

TP = tire penetration, in inches (may be different for the haul and the return)

EXAMPLE

Determine the rolling resistance (haul and return) for a CAT 621B scraper carrying a 20.5-ton load if the tire penetration during the haul is 3 inches and the tire penetration on the return is 1 inch.

$$RR_{Haul} = (40 + [30 \times 3 \text{ inches}]) \times 53.8 \text{ tons} = 6,994 \text{ pounds}$$

$$RR_{Return} = (40 + [30 \times 1 \text{ inch}]) \times 33.3 \text{ tons} = 2,331 \text{ pounds}$$

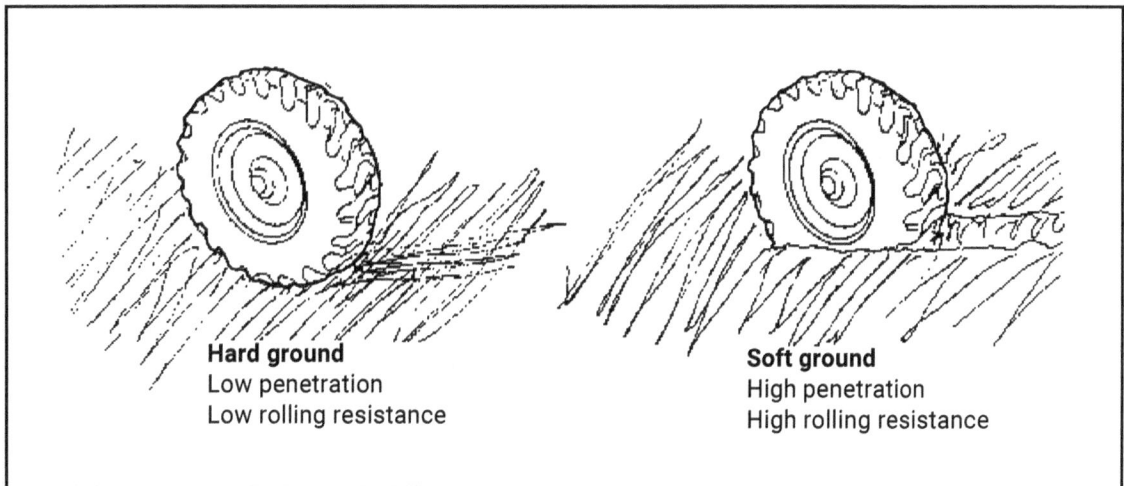

Figure 3-24. Rolling Resistance

Table 3-7. Representative Rolling-Resistance Values

Road Condition	Resistance Value (Pounds Per Ton)
Hard, smooth surface with no tire penetration (well maintained)	40
Firm, smooth surface, flexing slightly under load (well maintalned)	65

Flexible dirt roadway (irregular surface):
   With about 1 inch of tire penetration
   With up to 4 inches of tire penetration
	100
	150

Soft, muddy roadway (irregular surface or sand) with over 6 inches of tire penetration	220 to 400

Step 4. Determine the grade resistance or the grade assistance. Grade resistance is the opposing force of gravity that a vehicle must overcome to move uphill. Grade assistance is the helping force of gravity that pulls a vehicle downhill. For uphill (adverse) grades, the vehicle needs more power to move as it must overcome both rolling and grade resistance. For downhill (favorable) grades, the helping force of gravity produces additional pounds of pull to propel the vehicle. Indicate adverse grades by a plus (+) and favorable grades by a minus (-). In earthmoving, measure grades in percent of slope. This is the ratio between the vertical rise or fall, and the horizontal distance in which the rise or fall occurs. For instance, a rise of 1 foot in a 20-foot horizontal distance is a +5

percent grade $1/2$ &100(Use the following formula to determine the grade resistance or grade assistance:

GR(+) or GA(-) $= 20 \times$ **percent grade** $\times$ **vehicle weight (tons)**

Therefore—

GR(+) $_{Haul}$ **or GA(-)** $_{Haul}$ **2θpercent grade** $\times$ **GVW (tons)**

GR(+) $_{Return}$ **or GA(-)** $_{Return}$ $= 20 \times$ **percent grade** $\times$ **EVW (tons)**

where—
GR(+) = grade resistance, in pounds
GA(-) = grade assistance, in pounds
20 = constant that represents 20 pounds per ton of vehicle weight per degree of slope

NOTE: Enter the percent grade as a percent not as a decimal.

EXAMPLE

Determine the grade resistance and grade assistance for a CAT 621B scraper carrying a 20.5-ton load on a -5 percent grade from cut to fill.

GA(-) $_{Haul}$ **2θ(-5 percent)** $\times$ **53.8 tons (GVW)** $= $ **-5,380 pounds**

GR(+) $_{Return}$ **2θ(+5 percent)** $\times$ **33.3 tons (EVW)** $= $ **+3,330 pounds**

Step 5. Determine the rimpull required. Rimpull required is a measure of the force needed (in pounds) to overcome the vehicle's rolling resistance and grade assistance/grade resistance. Use the following formula to determine the rimpull required:

RPR $=$ **RR GA(-) or GR(+)**

where—
RPR = rimpull required, in pounds
RR = rolling resistance, in pounds
GA(-) = grade assistance, in pounds
GR(+) = grade resistance, in pounds

EXAMPLE

Determine the rimpull required on the haul and return based on the following data:

RR $_{Haul}$ **= 6,994 pounds; GA(-) = -5,380 pounds**

RR $_{Return}$ **= 2,331 pounds; GR(+) = +3,330 pounds**

RPR $_{Haul}$ **= 6,994 pounds +(-5,380) pounds = 1,614 pounds**

RPR $_{Return}$ **= 2,331 pounds +(+3,330) pounds = 5,661 pounds**

Available rimpull is the amount of force that can actually be developed as limited by traction. The engine may be able to develop the rimpull, but the rimpull must be able to be transferred at the point where the tire touches the ground. Therefore, required rimpull must always be less than available rimpull, or there will be tire slippage and the work will not be accomplished.

Step 6. Determine the travel speed.

- The travel speed of a piece of equipment is the maximum speed at which the vehicle can develop the rimpull required to overcome the opposing forces of grade and rolling resistance. The manufacturer normally provides this information in tables or charts. *Figures 3-8* and *3-9* show rimpull charts for the CAT 621B and the CAT 613B.

- To determine the travel speed, locate the rimpull required for either the haul or return on the left side of the chart. Read to the right until intersecting the line representing the highest gear which can achieve that amount of rimpull. Read down from the gear intersect to determine the maximum travel speed.

NOTE: Determine the travel speed for both the haul and the return.

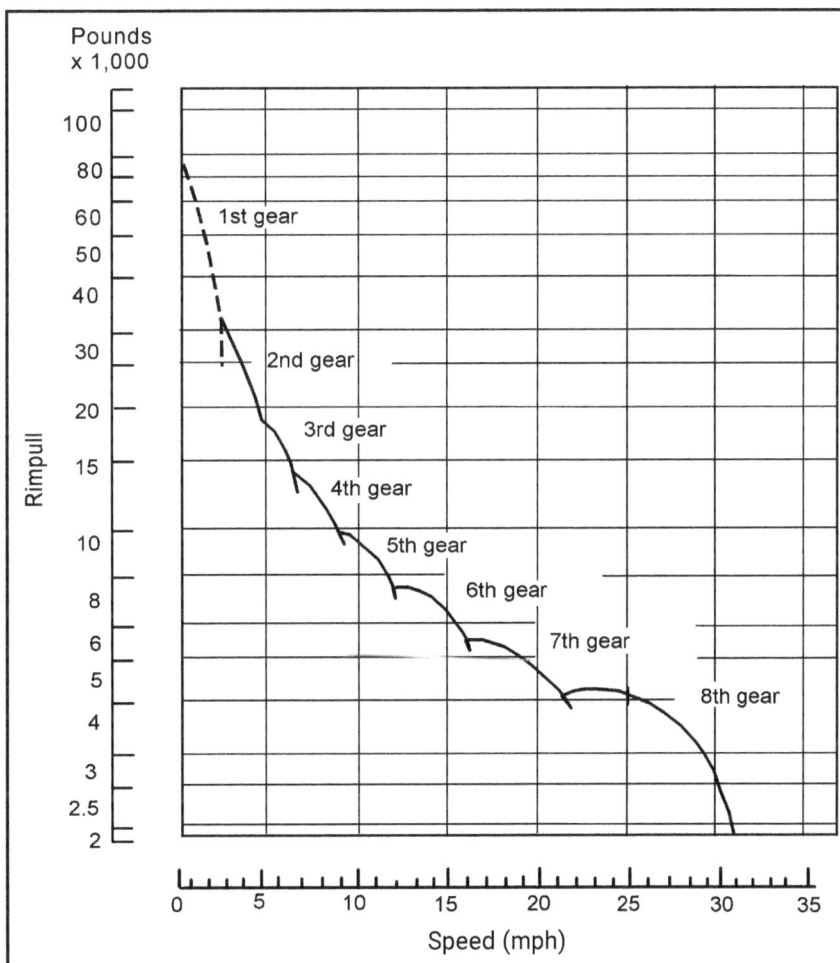

Figure 3-25. Speed Chart for the CAT 621B

Pounds
x 1,000

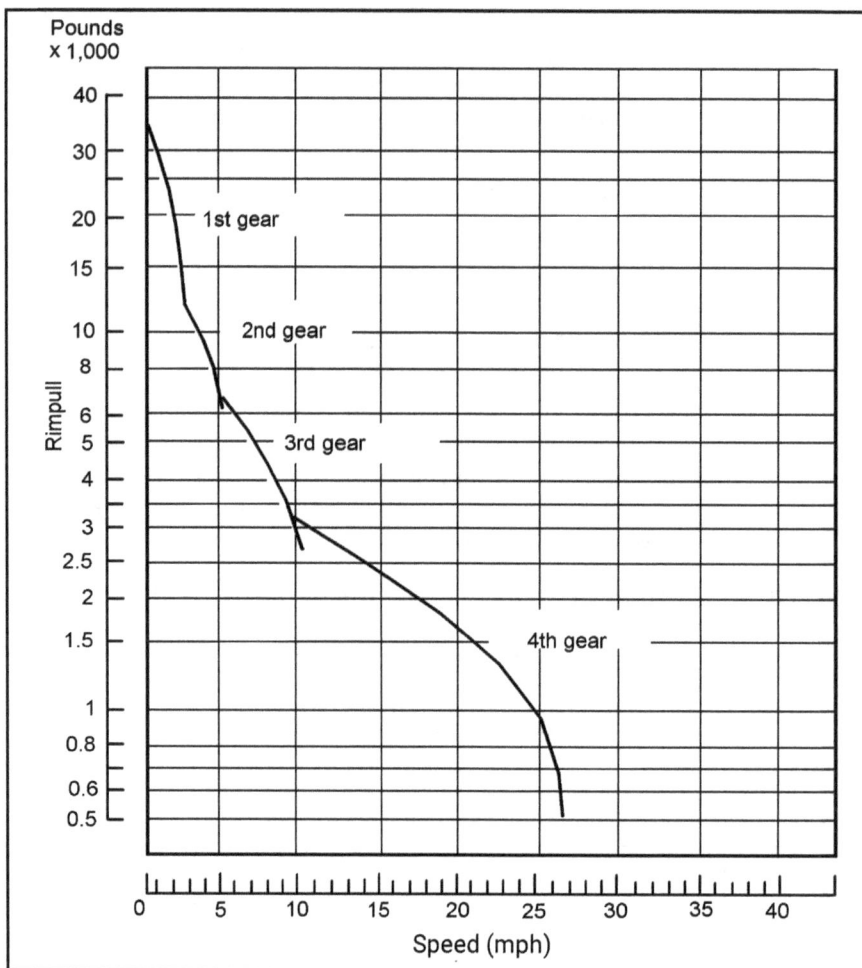

Figure 3-26. Speed Chart for the CAT 613B

EXAMPLE

Determine the maximum travel speed for a CAT 621B scraper, based on the following data.

RPR $_{Haul}$ = 1,614 pounds

RPR $_{Return}$ = 5,661 pounds

First, determine the travel speed for the haul. Refer to *Figure 3-8* and locate 1,614 pounds on the scale. This is below the lowest scale number of 2,000 pounds so use the bottom line on the rimpull scale. Read right to determine travel gear (eighth gear) and down to determine travel speed (31 miles per hour [mph]).

Second, determine the travel speed for the return. Refer to *Figure 3-8* and locate 5,661 pounds (interpolate between 5,000 and 6,000 on the rimpull scale). Read right to determine travel gear (seventh gear) and down to determine travel speed (17 mph).

NOTE: If computed travel speed (either haul or return) exceeds the unit's standing operating procedure (SOP) maximum allowable speed, determine the travel time based on the maximum allowable speed in the SOP.

Step 7. Determine the total travel time. Total travel time is the sum of the time it takes the vehicle to complete one haul and one return.

$$\text{Total TT} = TT_{\text{Haul}} + TT_{\text{Return}}$$

where—
   TT = travel time
- First, determine the haul travel time.

$$TT_{\text{Haul}} = \frac{\text{average haul distance (feet)}}{88 \times \text{travel speed}_{\text{Haul}}}$$

where—
   TT = travel time, in minutes
   88 = conversion factor used to convert mph into feet per minute (fpm)
- Second, determine the return travel time (in minutes).

$$TT_{\text{Return}} = \frac{\text{average return distance (feet)}}{88 \times \text{travel speed}_{\text{Return}}}$$

where—
   TT = travel time, in minutes
   88 = conversion factor used to convert mph into fpm

NOTE: The haul and return routes are not always the same. Be sure to use the correct haul distance for each computation.

EXAMPLE

Determine the total travel time for a CAT 621B based on a haul speed of 31 mph, a return speed of 17 mph, and a haul distance of 800 feet. The unit's SOP limits scraper travel speed to 25 mph.

Determine the haul and the return travel time.

$$TT_{\text{Haul}} = \frac{800 \text{ feet}}{88 \times 25 \text{ mph}} = 0.36 \text{ minute}$$

$$TT_{\text{Return}} = \frac{800 \text{ feet}}{88 \times 17 \text{ mph}} = 0.54 \text{ minute}$$

Determine the total travel time.

$$\text{Total TT} = 0.36 \text{ minute} + 0.54 \text{ minute} = 0.9 \text{ minute}$$

Step 8. Determine the cycle time.
- The cycle time is the sum of the total travel time and the time required for loading, dumping, turning at the dump site, and turning and positioning to load, plus the time to accelerate/decelerate during the haul and return.
- The average dump time for scrapers having a heaped capacity of less than 25 cubic yards is 0.3 minute. The type or size of the scraper does not significantly affect the turning time. Average turning time in the cut is 0.3 minute and 0.21 minute on the fill. The cut turning time is slightly higher because of congestion in the area and the necessity of spotting for loading. Therefore, for both the CAT 613 and the CAT 621 scrapers, allow 0.81 minute for dumping, turning at the dump site,

and turning at the load site. The question of the time for loading is the consequential variable.

- A good average time for loading the CAT 621 with a D8 or equivalent-size push tractor is 0.85 minute. Modify the time for loading or the assumed load volume if using a smaller push tractor. With a D7, expect load times approaching 1 minute. The self-loading CAT 613 requires 0.9 minute to load in good material. Good means loam, loose clay, or sandy material. Encountering tight materials will increase the loading duration. To determine the turn-and-dump time and the load time for a special piece of equipment, time the equipment as it goes through a few cycles.

$$CT = total\ TT + TD + LT$$

where—
CT = total scraper cycle time, in minutes
TT = travel time
TD = total turn and dump time
LT = load time

EXAMPLE

Determine the cycle time for a CAT 621B scraper with a D7 push tractor based on a travel time of 0.9 minute and an average turn and dump and load time.

CT (minutes) = 0.9 minute + 0.81 minute + 1 minute = 2.71 minutes

---

Step 9. Determine the trips per hour. To determine the number of trips per hour, divide the working minutes per hour by the cycle time. Normally there are about 50 minutes per hour of productive time on a well-managed scraper job. However, if the cut is in a tight area such as a ditch or if the embankment is a narrow bridge header, the estimator should consider lowering the productive time to a 45-minute working hour.

$$TPH = \frac{working\ minutes\ per\ hour}{CT}$$

where—
TPH = trips per hour
CT = total scraper cycle time, in minutes

EXAMPLE

Determine how many trips per hour a CAT 621B can make based on a 50-minute working hour and a cycle time of 2.71 minutes per trip.

$$TPH = \frac{50\ minutes}{2.71\ minutes\ per\ trip} = 18.5\ trips$$

Step 10. Determine the hourly production rate. To determine the hourly production rate, the average size of the load (in LCY) and the number of trips per hour must be known. The capacity of the scraper, the material type, and the method of loading will determine the average size of load.

$$P = TPH \times \text{average LCY per load}$$

where—
P = hourly production rate, in LCY per hour
TPH = trips per hour

NOTE: To convert from LCY to either BCY or CCY, multiply the production rate by a soil conversion factor from *Table 1-1, page 1-4.*

$$P \text{ (BCY per hour or CCY per hour)} = P \text{(LCY per hour)} \times \text{conversion factor}$$

where—
P = hourly production rate

EXAMPLE

Determine the hourly production rate in BCY per hour for a CAT 621B working in loam, making 18.5 trips per hour, with an average load of 20 LCY.

$$P \text{ (LCY per hour)} = 18.5 \text{ TPH} \times 20 \text{ LCY per load} = 370 \text{ LCY per hour}$$

$$P \text{ (BCY per hour)} = 370 \text{ LCY per hour} \times 0.8 = 296 \text{ BCY per hour}$$

Step 11. Determine the total time in hours required to complete the mission. To determine the total time required to complete a mission, the total volume to move, the hourly production rate, and the number of scrapers to be used on the job must be known.

$$\text{Total time (hours)} = \frac{Q}{P \times N}$$

where—
Q = total volume to move, in BCY
P = hourly production rate, in BCY per hour
N = number of scrapers

EXAMPLE

Determine how many hours it would take to move 19,440 BCY, using three CAT 621B scrapers, each with an hourly production rate of 296 BCY per hour.

$$\text{Total time (hours)} = \frac{19,440}{296 \text{ BCY per hour} \times 3} = 22 \text{ hours}$$

If it is necessary to complete the job in a specified time, use the same basic formula to determine the required number of scrapers.

$$\text{Number of scrapers required} = \frac{Q}{P \times H}$$

where—
> Q = total volume to move, in BCY
> P = hourly production rate, in BCY per hour
> H = required number of hours

Step 12. Determine the number of push tractors required. The number of push tractors required is a ratio of the scraper cycle time to the push-tractor cycle time. The self-loading CAT 613 does not use a push tractor, so this part of the analysis is not necessary when using self-loading scrapers.

$$N = \frac{CT}{PT}$$

where—
> N = number of push tractors required
> CT = total scraper cycle time, in minutes
> PT = total pusher cycle time, in minutes

- Load time (discussed in *step 8*). A CAT 621B loading with a D7 push tractor requires about 1 minute to load. This is the time the push tractor is in contact with the scraper.
- Push-tractor cycle time. Once a scraper load time has been determined, use the following formula to determine the push-tractor cycle time.

$$PT = (1.4 \times LT) + 0.25$$

where—
> PT = total push-tractor cycle time
> 1.4 = constant that represents scraper load time and push-tractor travel time between scrapers
> LT = load time
> 0.25 = constant that represents push-tractor positioning time

At this point, the number of scrapers that a single push tractor will support can be determined.

EXAMPLE

Determine how many CAT 621B scrapers a single push tractor can support if the scraper cycle time is 2.71 minutes and the scraper load time is 1 minute.

$$PT = (1.4 \times 1) + 0.25 = 1.65 \text{ minutes}$$

$$\text{Number of scrapers} = \frac{2.71 \text{ minutes}}{1.65 \text{ minutes}} = 1.64 \text{ scrapers}$$

This example shows that the push-tractor cycle time will control the production when using only one push tractor and more than one scraper on the project. The push-tractor production formula is—

$$P = \frac{\text{working minutes per hour}}{\text{PT}} \times \text{scraper load (LCY)}$$

where—
P = hourly production rate, in LCY per hour
PT = total push-tractor cycle time, in minutes

As was done in step 10, convert the production into BCY or CCY by using the *Table 1-1, page 1-4,* soil conversion factors.

EXAMPLE

Determine what the production will be in BCY if a single push tractor, with a cycle time of 1.65 minutes supports two CAT 621B scrapers hauling 20 LCY of loam. Assume a 50-minute working hour. The scrapers have a cycle time of 2.71 minutes.

$$\text{Number of scrapers one pusher can support} = \frac{2.71 \text{ minutes}}{1.65 \text{ minutes}} = 1.64 \text{ scrapers}$$

Therefore, if using only one push tractor, the pusher cycle time will control production.

$$P \text{ (BCY per hour)} = \frac{50}{1.65} \times 20 \text{ LCY} \times 0.8 = 485 \text{ BCY per hour}$$

NOTE: If the incorrect assumption was made that one pusher could handle two scrapers, the production would have been calculated at 590 BCY per hour.

$$P \text{ (BCY per hour)} = \frac{50 \text{ minutes}}{2.71 \text{ minutes}} \times 2 \text{ scrapers} \times 20 \text{ LCY} \times 0.8 = 590 \text{ BCY per hour}$$

where—
P = hourly production rate

Once the number of scrapers that one push tractor can support has been determined, use the following formula to determine how many push tractors are needed to support the job if using additional scrapers.

$$\text{Number of push tractors required} = \frac{\text{number of scrapers on job}}{\text{number of scrapers a push tractor can support}}$$

EXAMPLE

Determine how many push tractors are required on a job that has nine 621B scrapers, if a single push tractor can support 1.64 scrapers.

$$\text{Number of push tractors required} = \frac{9 \text{ scrapers}}{1.64 \text{ scrapers per push tractor}} = 6 \text{ push tractors}$$

Chapter 4

Graders

Graders are multipurpose machines used for grading, shaping, bank sloping, and ditching. They are used for mixing, spreading, side casting, leveling and crowning, general construction, and road and runway maintenance. Graders cannot perform dozer work because of the structural strength and location of its blade. However, they can move small amounts of material. They are capable of working on slopes as steep as 3:1. Graders are capable of progressively cutting ditches to a depth of 3 feet.

GRADER COMPONENTS

4-1. The components of the grader that do the work are the blade and the scarifier. The blade's position and pitch are adjustable and are determined by the type of operation being performed.

BLADE

4-2. The major component of a grader blade is a hydraulically controlled moldboard to which the cutting edges are bolted. Use the blade *(Figure 4-1, page 4-2)* to side cast material. The ends of the blade can be raised or lowered together or independently of one another.

Blade Position

4-3. The blade can be angled perpendicular to the line of travel or parallel to the direction of travel. It can also be shifted to either side or raised into a vertical position *(Figure 4-2, page 4-3)*.

Blade Pitch

4-4. The blade can be pitched forward or backward *(Figure 4-3, page 4-3)*. Keep the blade near the center of the pitch adjustment; this keeps the top of the moldboard directly over the cutting edge of the blade. Pitching the blade forward decreases the blade's cutting ability and increases the dragging action. The blade will tend to ride over the material rather than cut and push, and it has less chance of catching on solid obstructions. Use a forward pitch to make light, rapid cuts and to blend materials. When the blade is pitched to the rear, it cuts readily but the material tends to boil over itself.

SCARIFIER

4-5. Use a scarifier (see *Figure 4-1*) to break up material too hard for the blade to cut. A scarifier has 11 removable teeth that can be adjusted to cut a maximum depth of 12 inches. When operating in hard material, it may be necessary to remove some of the teeth from the scarifier. Do not remove more than five teeth because the force against the remaining teeth could shear them off. When removing teeth, take the center one out first and then alternately remove the other four teeth. This balances the scarifier and distributes the load evenly. With the top of the scarifier pitched to the rear, the teeth lift and tear the material being loosened. Use this position for breaking up asphalt pavement. Adjust the pitch of the scarifier for the type of material being ripped.

ROAD AND DITCH CONSTRUCTION

4-6. Road grading, embankment finish work, and shallow-ditch construction are basic grader operations.

MARKING FOR A DITCH CUT

4-7. For better grader control and straighter ditches, make a 3- to 4-inch-deep marking cut on the first pass *(Figure 4-4, page 4-4)* at the outer edge of the bank slope (usually identified by slope stakes). The toe of the blade should be in line with the outside edge of the lead wheel. This marking cut provides a guide for subsequent operations.

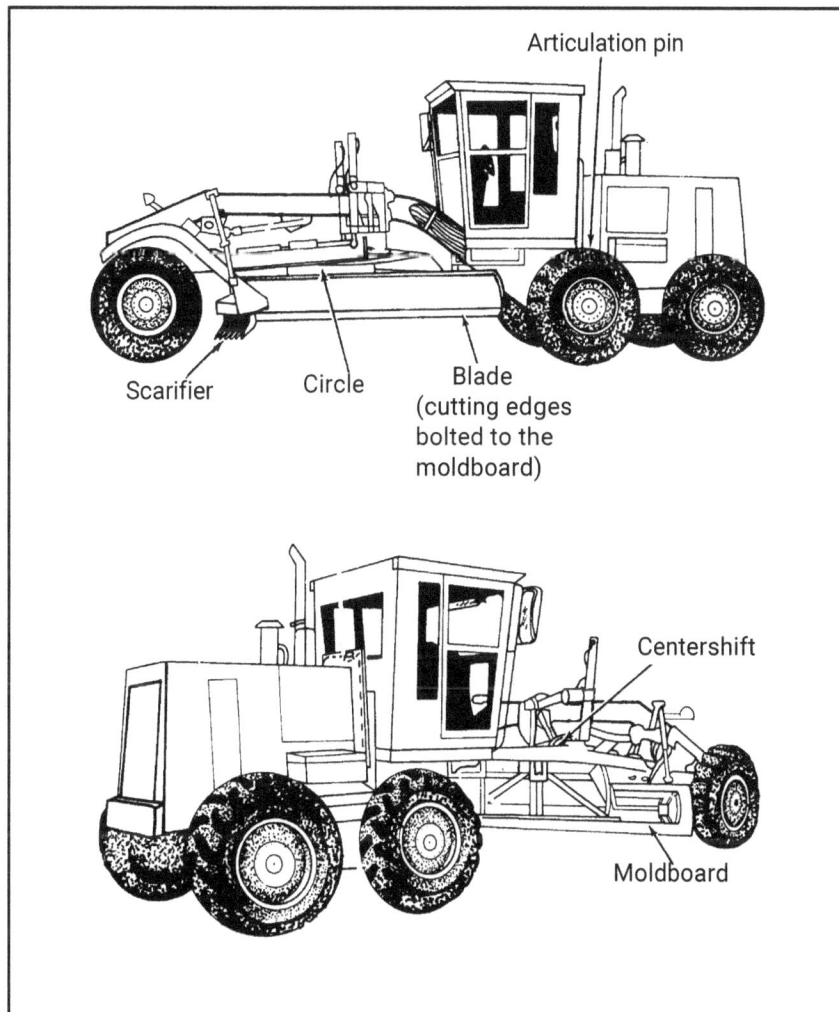

Figure 4-1. Grader

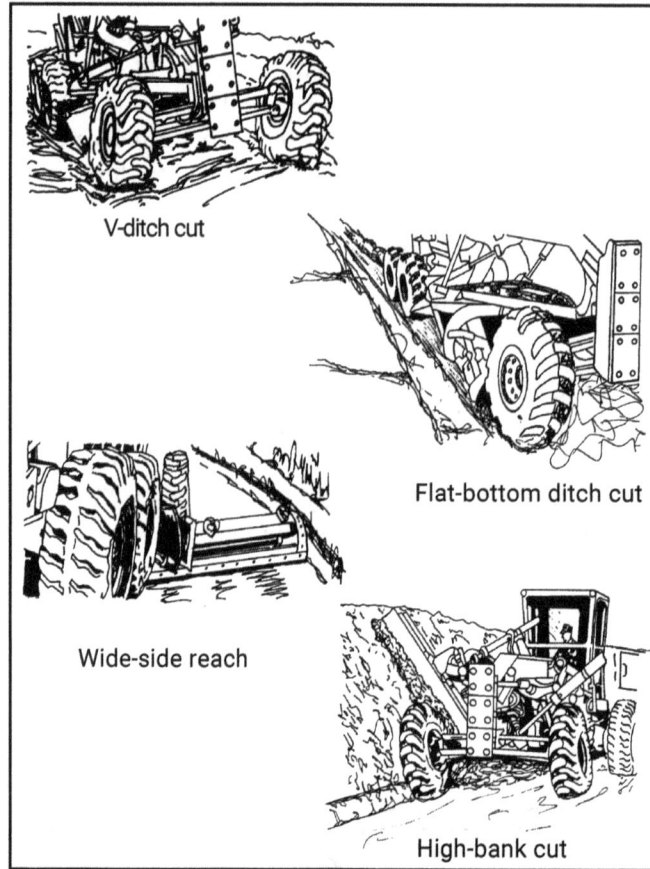

Figure 4-2. Blade Positions

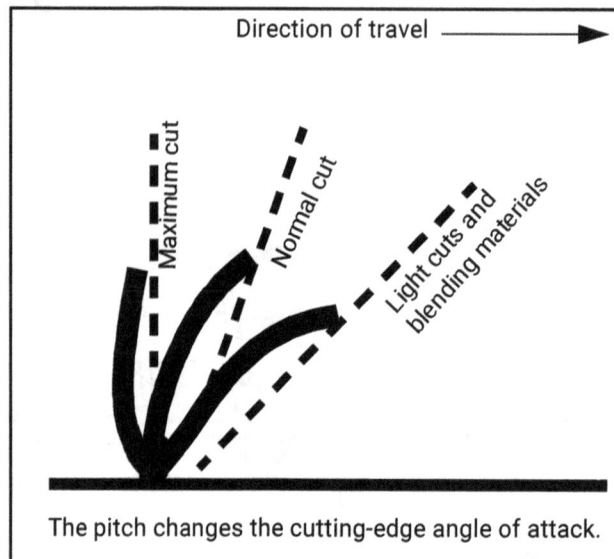

Figure 4-3. Blade Pitch

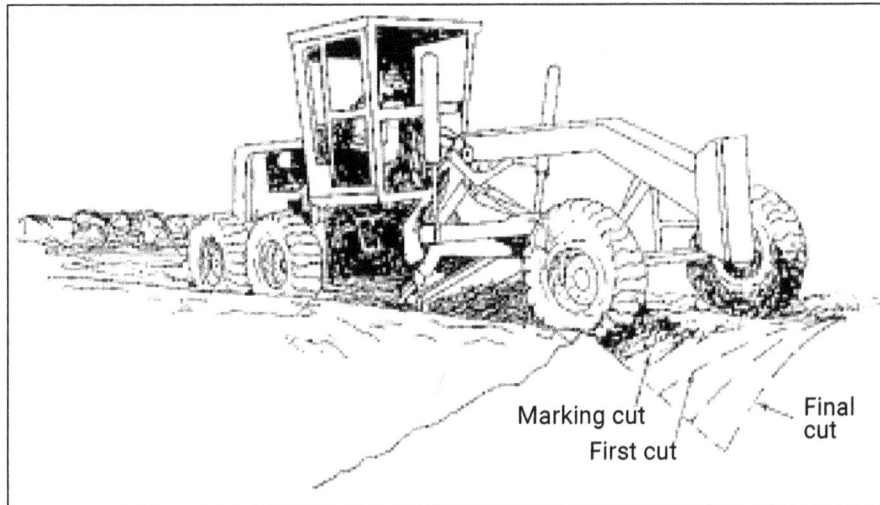

Figure 4-4. Starting a Ditch

MAKING A DITCH CUT

4-8. Make each ditch cut as deep as possible without stalling or losing control of the grader. Normally, make ditching cuts in second gear at full throttle. Start with the blade positioned so that the toe is in line with the center of the lead wheel. Bring each successive cut in from the edge of the bank slope so that the toe of the blade will be in line with the ditch bottom on the final cut. *Figure 4-5* shows the steps of the V-ditching method. The steps shown in *Figure 4-5* are for a single roadside ditch. Repeat the steps on the opposite side of the road. The machine's frame should be articulated when performing steps 4 and 7.

Marking the Cut

Step 1. Begin the ditch by establishing a marking cut as follows (ditching is normally done on the right-hand side of the grader):

- Ensure that the moldboard is high enough off the surface to allow unrestricted movement.
- Ensure that the blade is pitched halfway.
- Center shift until the left lift cylinder (heel) is straight up and down.
- Rotate the moldboard so that the toe is just behind the outside edge of the right front wheel (about a 45° angle to the frame).
- Side shift the blade if necessary to extend the edge of the moldboard to the outside edge of the right front wheel.
- Raise the left lift cylinder all the way.
- Lean the front wheels to the left. The grader is now in the ditching position.

Step 2. Move the grader forward. As the right front wheel passes over the starting point of the ditch, lower the toe of the moldboard. Apply enough pressure on the toe to penetrate the ground's surface about 3 to 4 inches.

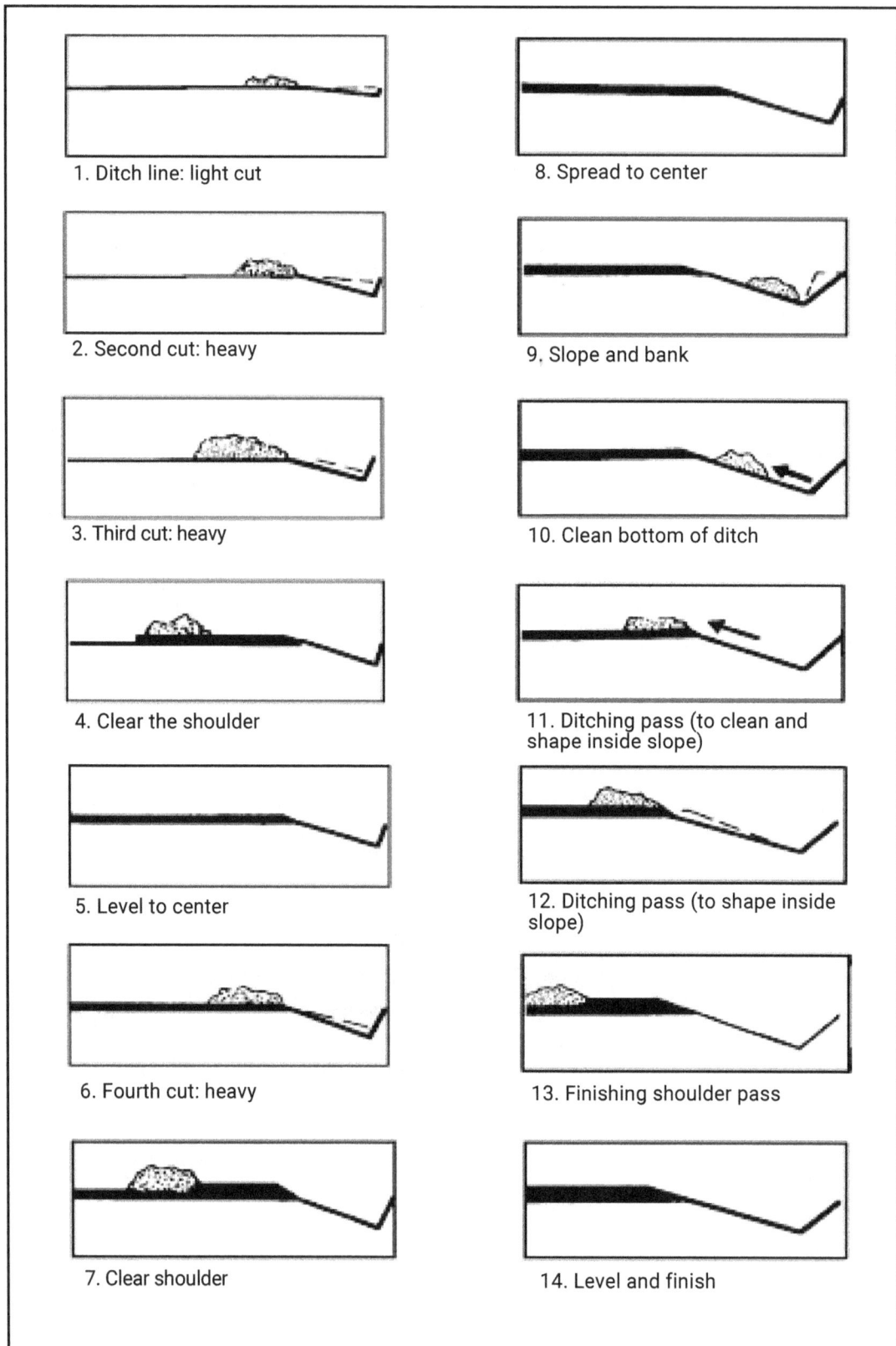

Figure 4-5. V-ditching Method

1. Ditch line: light cut
2. Second cut: heavy
3. Third cut: heavy
4. Clear the shoulder
5. Level to center
6. Fourth cut: heavy
7. Clear shoulder
8. Spread to center
9. Slope and bank
10. Clean bottom of ditch
11. Ditching pass (to clean and shape inside slope)
12. Ditching pass (to shape inside slope)
13. Finishing shoulder pass
14. Level and finish

Step 3. Feather the material and raise the moldboard toe clear of the ground at the completion of the marking cut. Continue moving forward until the rear wheels pass over and off the marking cut.

NOTE: Feathering is accomplished by raising the moldboard in 1/2- to 1-inch increments while moving forward. Two or three seconds are recommended between each upward adjustment until all the material in front of the moldboard passes under it.

Step 4. Straighten the front wheels and steer the grader to the right (about a 45° angle to the ditch).

Step 5. Back the grader along the outside edge of the windrow.

Step 6. Reposition the grader at the start point.

Step 7. Lean the front wheels to the left.

Establishing the Depth of the Ditch

Step 1. Place the grader in forward motion and apply as much downward pressure to the toe of the blade that the grader will handle.

Step 2. Continue along the ditch line until the grader has reached the finishing point, and then follow the exit procedures previously discussed under marking the cut.

NOTE: When making ditch cuts, windrows form between the heel of the blade and the left rear wheel. Move or level these windrows when either the ditch is at the planned depth or the windrow becomes higher than the road clearance of the grader. This material will form the shoulder of the road.

Establishing the Shoulder of the Ditch

4-9. This task is accomplished by placing the grader in the wide-side reach position.

Step 1. Adjust the moldboard as follows:
- Rotate the moldboard to a 90° angle (perpendicular) with the frame (straight across) and adjust the height of the blade to about 4 to 6 inches above the surface.
- Center shift the blade all the way to the right.
- Readjust the height of the blade to about 2 inches above the surface.
- Side shift the blade all the way to the right.
- Lean the front wheels to the left.
- Circle the moldboard counterclockwise until the toe is about 12 to 15 inches from the outside edge of the front right wheel.

NOTE: Do not adjust the moldboard height, especially the left lift cylinder.

Step 2. Move the grader forward and maintain a position and course so that the toe of the moldboard passes directly over the center of the ditch.

Step 3. Apply enough downward pressure to skim the material from the shoulder; do not cut the shoulder.

Step 4. Continue forward as the grader passes the finishing point of the ditch until all the material in front of the moldboard passes under it or is windrowed off the heel.

Step 5. Continue forward until enough space is available to position the grader to back up and straddle the windrow.

NOTE: Place the grader in the right-hand general grade position and the moldboard will be positioned to execute the next maneuver. Do not back the grader in the wide-side reach position.

Step 6. Ensure that the front wheels are straight up and down before backing the grader.

Step 7. Back the grader to the starting point of the project and, after stopping, lean the wheels to the left.

Step 8. Lower the toe and heel of the moldboard to the surface.

Step 9. Raise the heel about 2 to 3 inches and ensure that the toe is just touching the surface. With the heel raised about 3 inches, the loose material from the ditch should pass under and off the heel of the moldboard.

Step 10. Move the grader forward. Maintain a straight course by keeping the grader centered on the windrow.

Step 11. Skim the shoulder of the road with the toe and spread the windrow to form the surface of the road.

Step 12. Ensure that the material is feathered at the end of the pass before stopping the grader.

Step 13. Straighten the front wheels and raise both lift cylinders all the way.

Step 14. Reposition the grader at the finishing end of the project. The grader should be positioned to establish a V-ditch (going the opposite direction) on the other side of the project area.

NOTE: Sometimes ditch cuts produce more material than is needed for the roadbed and shoulders. Use this excess material as fill at other locations throughout the project. Blade the excess material into a windrow and haul it to the appropriate location.

CREATING A BANK SLOPE

4-10. Sloping the bank on a road cut prevents slope-sloughing failures. It also prevents excessive erosion of the bank, which could fill the roadside ditch. Initially, cast the material cut from the outer slope into the bottom of the ditch and remove it later. *Figure 4-6, page 4-8,* shows a grader sloping a high-bank cut.

CLEANING A DITCH

4-11. To remove unwanted material that was pushed into the ditch during the bank slope operation, place the blade in the same position as used for the ditching cuts. This casts the material onto the shoulder.

FINISHING A SHOULDER

4-12. Move the windrow (formed by cleaning the ditch) onto the road at the same time the shoulder is being finished to the desired slope.

FINISHING A CROWN OR A CROSS SLOPE

4-13. The final operation is to spread all the material brought from the ditch onto the roadway. Use this material to bring the roadway to the desired crown or a cross slope.

For heavier cut, lean wheels toward slope.
For lighter cut, lean wheels away from slope.

Figure 4-6. Sloping a High Bank

EARTH- AND GRAVEL-ROAD MAINTENANCE

LEVELING AND MAINTAINING SURFACES

4-14. Ordinarily, level and maintain a surface by working the material across the road or runway from one side to the other. However, to maintain a satisfactory surface in dry weather, work traffic-eroded material from the edges and shoulders of the road toward the center. Traffic or wind can cause loss of binder material, so be cautious when disturbing dry road surfaces. The surface is easier to work if it is damp; therefore, after a rain is a good time to perform surface maintenance. A water truck may be necessary to dampen dry material.

Step 1. Rotate the moldboard so that the toe is on the right side of the grader at about a 50° to 60° angle to the frame.

Step 2. Ensure that the blade is pitched halfway.

Step 3. Center shift the blade until the left lift cylinder is straight up and down.

Step 4. Lean the front wheels to the left.

Step 5. Lower the moldboard until the toe and heel slightly touch the ground.

Step 6. Place the grader in motion and, as the moldboard crosses the project start line, apply enough downward pressure on both the heel and the toe to penetrate the surface on a level plain about 1/2 inch.

Step 7. Maintain a straight course, adjusting the moldboard slightly to carry the material the length of the project.

Step 8. Feather the material at the end of the pass.

Step 9. Stop the grader and straighten the front wheels after the material is feathered to a smooth termination.

Step 10. Raise both lift cylinders all the way.

Step 11. Position the grader to straddle the windrow just made, and back the grader to the starting point while ensuring the windrow is between the wheels (do not drive on top of the windrow).

Step 12. Stop the grader just outside of the project boundary line.

Repeat this process until the entire area is leveled.

SMOOTHING PITTED SURFACES

4-15. When binder is present and moisture content is appropriate, rough or badly-pitted surfaces may be cut smooth. The cut surface material is then respread over the smooth base. Again, the best time to reshape earth and gravel roads is after a rain. Dry roads should be moistened by using a water distributor. This ensures that the material will have sufficient moisture content to recompact readily.

CORRECTING CORRUGATED ROADS

4-16. When correcting corrugated roads, be careful not to make the situation worse. Deep cuts on a washboard surface will set up blade chatter, which emphasizes rather than corrects corrugations. Badly-corrugated surfaces may require scarifying. With proper moisture content, level the surface by cutting across the corrugations. Alternate the blade angle so that the cutting edge will not follow the rough surface. Cut the surface to the bottom of the corrugations, and then reshape the surface by spreading the windrows in an even layer across the road. After reshaping the road, the traffic will compact it. However, rolling the surface after shaping will give better and longer-lasting results.

SCARIFYING ROADS OR AREAS

Step 1. Position the grader outside the working area.

Step 2. Ensure that the moldboard is high enough off the ground to allow unrestricted movement.

Step 3. Rotate the moldboard so that it is perpendicular with the frame, and adjust the height to 12 inches off the surface (level).

Step 4. Center shift the blade until the lift cylinders are centered on the grader.

Step 5. Pitch the blade halfway.

Step 6. Ensure that the front wheels are vertical.

Step 7. Move the grader forward.

Step 8. Lower the scarifier as it crosses the starting point and penetrate the surface.

Step 9. Scarify the entire length of the area to a minimum depth of about 6 inches.

Step 10. Raise the scarifier at the finish point.

Step 11. Exit the project area and stop the grader.

Step 12. Rotate the moldboard to a 50° angle, and adjust the center shift to straighten the heel cylinder.

Step 13. Return to the starting point, and reposition the grader for a second scarifying pass.

SNOW REMOVAL

4-17. Graders remove snow in much the same way as snowplows. Be sure to raise the blade 0.5 to 1 inch when removing snow from uneven pavements or portable runway surfaces. Improper adjustment can damage the grader and gouge the surface.

ASPHALT MIXING

4-18. Asphalt can be mixed in place or mixed with imported aggregate. *Chapter 12* provides additional information on asphalt mixing.

MIXED-IN-PLACE ASPHALT

4-19. For mixed-in-place asphalt, spread the asphalt directly on the road surface, either before or after scarifying the surface. After applying the asphalt, mix it with the surface soil by scarifying and/or windrowing with the blade.

IMPORTED AGGREGATE

4-20. When using imported aggregate for a pavement—

Step 1. Shape the existing base and prepare it by blading, rolling, and curing as necessary.

Step 2. Dump the aggregate mix and blade it into uniform windrows. If the aggregate is too wet, blade the windrows to allow evaporation of the excess moisture.

Step 3. Flatten the windrow and apply the asphalt.

Step 4. Mix the asphalt with the aggregate using the grader. Move the windrow from side to side across the road by making successive passes with the blade. Several graders can operate, one behind another (tandem), on the same windrow. If rain moistens the mixture, continue mixing until dry.

Step 5. Blade the material back into a windrow after mixing and before spreading.

LARGE-AREA MIXTURES

4-21. Set stakes to mark the edges of the spread width for each windrow. When spreading mixtures over large areas, drive blue-top hubs (blue tops) to indicate final pavement elevation. The blue tops are usually placed in a grid pattern 20 feet apart. Remove the blue tops before rolling the pavement. Usually, one pass of the grader will flatten the windrow after which it can be spread to each side in increments. This produces a layer of uniform thickness with proper lateral and longitudinal slopes. A skilled grader operator is essential at this phase.

OPERATION TECHNIQUES AND TIPS

4-22. Graders can be used for spreading, leveling, side casting, and planing materials. Different procedures are required to achieve a desired result.

SPREADING AND LEVELING

4-23. Use a grader to spread windrows of loose material *(Figure 4-7)*. If there is space to work around the sides of the windrows, extend the blade well to the side and reduce the windrow, using a series of side cuts. Spread the windrows as much as possible. The power and traction of the grader will limit the load to be pushed. Graders have less power and traction than dozers, but graders move the load faster. Although the grader blade is low, it is more concave than the dozer blade. This gives increased rolling action to the load so that a large quantity can be pushed without spilling over the top. Leveling large windrows may require two or more sidecuts with a grader*(Figure 4-8)*.

Figure 4-7. Spreading Windrowed Material

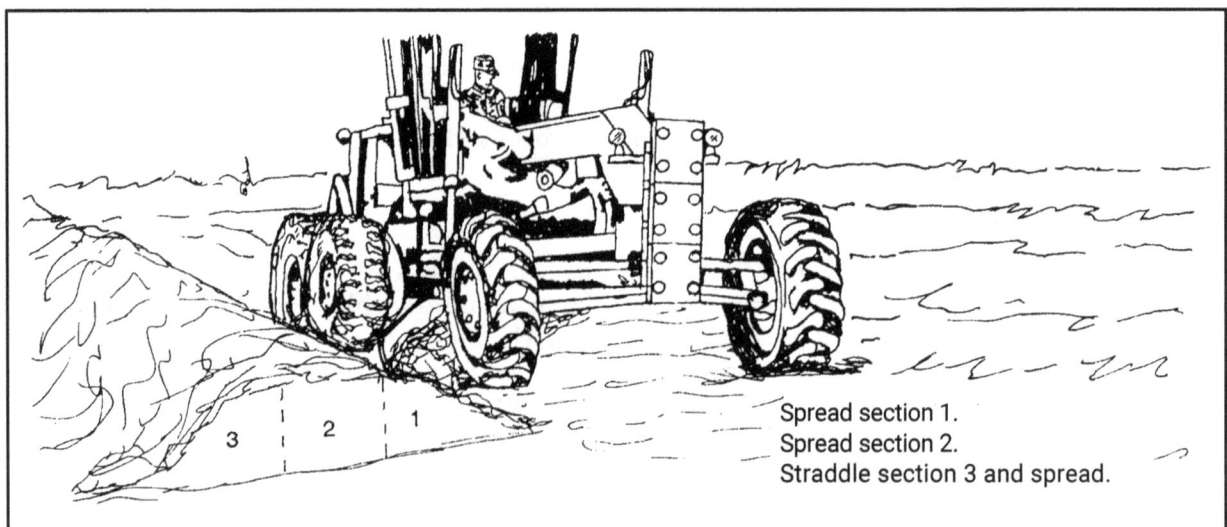

Spread section 1.
Spread section 2.
Straddle section 3 and spread.

Figure 4-8. Leveling Large Windrows

SIDE CASTING

4-24. Set the blade at an angle so that the load being pushed will drift off the trailing end *(Figure 4-9)*. Rolling action caused by the blade curve assists this side movement. As the blade is angled more sharply, the speed of the side drift increases (which does not carry the material as far forward) and deeper cuts can be made. To shape and maintain most roads, set the blade at a 25° to 30° angle. Decrease the angle for spreading windrows; increase the angle for hard cuts and ditching.

NOTE: A blade that is angled straight across (perpendicular to the direction of movement) is at 0°.

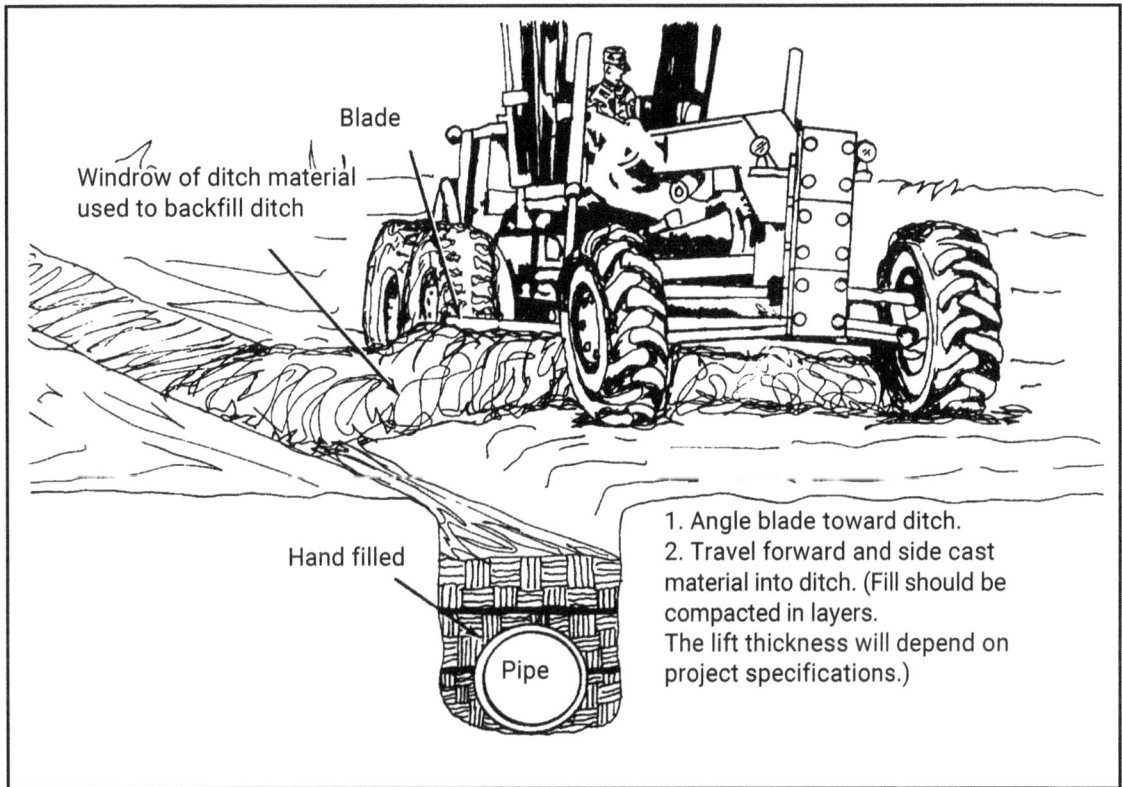

Figure 4-9. Backfilling a Ditch by Side Casting

PLANING

4-25. Set the blade at an angle to plane off irregular surfaces; use that material to fill the hollows. Cut enough material to always keep some in front of the blade. Move the loosened material forward and sideward to distribute it evenly. On the next pass, pick up the windrow that was left at the trailing edge of the blade. On the final pass, make a lighter cut and lift the trailing edge of the blade enough to allow the surplus material to go under rather than around the end. This will avoid leaving a ridge. Do not pile windrows in front of the rear wheels because it will adversely affect traction and grader control.

WORKING SPEEDS

4-26. Always operate the grader as fast as the operator's skill and the road conditions permit. Operate at full throttle in each gear. Use a lower gear if less speed is required, rather than operate at less than full throttle. *Table 4-1* lists the proper gear ranges for various grader operations under normal conditions. *Table 4-2* lists the road speeds for the Army's 130G grader.

Table 4-1. Proper Gear Ranges for Grader Operations

Operation	Gear
Maintenance	Second to third
Spreading	Third to fourth
Mixing	Fourth to sixth
Ditching	First to second
Bank sloping	First
Snow removal	Fifth to sixth
Finishing	Second to fourth

Table 4-2. Road Speeds for the Army's 130G Grader

| | Road Speed: mph at Rated rpm | | | | | | | |
| | Forward Gears | | | | | | Reverse Gears | |
Model	First	Second	Third	Fourth	Fifth	Sixth	Low	High
130G	2.3	3.7	5.9	9.7	15.5	24.5	Same as forward	

BLADE SETTING AND GRADER SPEED

4-27. Each job requires a specific blade setting and grader speed for optimum production. Deviations from these settings and speeds will cause machine inefficiency.

TURNING

4-28. When making a number of passes over a short distance (less than 1,000 feet), backing the grader to the starting point is normally more efficient than turning it around and continuing the work from the far end. Never make turns on a newly-laid bituminous road or runway surface.

NUMBER OF PASSES

4-29. Grader efficiency is directly proportional to the number of passes made. Operator skill, coupled with planning, is most important in eliminating unnecessary passes. For example, if four passes will complete a job, every additional pass increases the time and cost of the job.

TIRE INFLATION

4-30. Keep the tires properly inflated to get the best results. Overinflated tires result in less contact between the tires and the road surface, causing a loss of traction. Air-pressure differences in the rear tires cause tire slippage and grader bucking. The operator's manual gives the correct tire inflation pressure.

WET AND MUDDY CONDITIONS

4-31. Wet and muddy conditions cause poor traction, which may decrease grader efficiency. However, in spite of reduced efficiency, the grader is the best machine to use under these conditions. One example of this would be casting surface mud to the side on a haul road.

HAUL-ROAD MAINTENANCE

4-32. Keep haul roads in good condition. This will increase the efficiency of scrapers or dump trucks on large earthmoving operations. Graders are the best machines for maintaining haul roads. The most efficient method of road maintenance is to use enough graders to complete one side of a road with one pass of each grader (tandem operation). In this method, maintenance of one side of the road is completed while the other side is open to traffic.

TANDEM OPERATIONS

4-33. Using graders in tandem expedites such operations as leveling, mixing, spreading, and haul-road maintenance.

PRODUCTION ESTIMATES

4-34. Use the following formula to prepare estimates of the total time (in hours or minutes) required to complete a grader operation.

$$\text{Total time} = \frac{P \times D}{S \times E}$$

where—
  P = number of passes required
  D = distance traveled in each pass, in miles or feet
  S = speed of grader, in mph or fpm (multiply mph by 88 to convert to fpm)
  E = efficiency factor

- **Number of passes.** Estimate the number of passes (based on the project requirements) before construction begins.
- **Distance traveled.** Determine the required travel distance per pass before construction begins.
- **Speed of the grader.** Speed is the most difficult factor in the formula to estimate accurately. As work progresses, conditions may require that speed estimates be increased or decreased. Compute the work output for each rate of speed used in an operation. The speed depends largely on the skill of the operator and the material type.
- **Efficiency factor.** For grader operations the efficiency factor is usually no better than 60 percent.

EXAMPLE

Time estimate based on the number of miles of construction.

Maintenance of a 5-mile gravel road requires cleaning the ditches and leveling and reshaping the road. Use a CAT 130G grader and a 0.6 efficiency factor. Cleaning the ditches requires two passes in first gear, leveling the road requires two passes in second gear, and final shaping of the road requires three passes in fourth gear.

Speeds (from *Table 4-2, page 4-13*):
   First gear = 2.3 mph
   Second gear = 3.7 mph
   Fourth gear = 9.7 mph

$$\textbf{Total time} = \frac{2 \times 5}{2.3 \times 0.6} + \frac{2 \times 5}{3.7 \times 0.6} + \frac{3 \times 5}{9.7 \times 0.6} = 7.3 + 4.5 + 2.6 = \textbf{14.4 hours}$$

EXAMPLE

Time estimate based on the number of feet of construction.

A 1,500-foot gravel road requires leveling and reshaping. Use a CAT 130G grader with a 0.6 efficiency factor. The work requires two passes in second gear and three passes in third gear.

Speeds (from *Table 4-2*):
   Second gear = 3.7 mph
   Third gear = 5.9 mph

$$\textbf{Total time} = \frac{2 \times 1,500}{(88 \times 3.7) \times 0.6} + \frac{8 \times 1,500}{(88 \times 5.9) \times 0.6} = 15.4 + 14.4 = \textbf{29.8 minutes}$$

SAFETY

4-35. Listed below are specific safety rules for grader operators:

- Always display a red flag or a flashing light on a staff at least 6 feet above the left rear wheel when operating a grader slowly on a highway or roadway.
- Never allow other personnel to ride on the blade or rear of the grader.
- Always engage the clutch gently, especially when going uphill or pulling out of a ditch.
- Always reduce speed before making a turn or applying the brakes.
- Always keep the grader in low gear when going down steep slopes.
- Always take extra care when working on hillsides to drive slowly and to be observant of holes or ditches.
- Never use graders to pull stumps or other heavy loads.
- Always keep the blade angled well under the machine when it is not in use.
- Never allow more than one person on a grader while it is in operation. If it has a buddy seat, ensure that no more than two people are on the machine while it is in operation.

Chapter 5

Loaders

Loaders are used extensively in construction operations to handle and transport material, to load haul units, to excavate, and to charge aggregate bins at both asphalt and concrete plants. The loader is a versatile piece of equipment designed to excavate at or above wheel or track level. The hydraulic-activated lifting system exerts maximum breakout force with an upward motion of the bucket. Large rubber tires on wheel models provide good traction and low ground-bearing pressure. A wheel loader can attain high speeds, which permits it to travel from one job site to another under its own power.

DESCRIPTION

5-1. Military loaders are diesel-driven, rubber-tired machines *(Figure 5-1, page 5-2)*. They are available in varied sizes and capacities. A power-shift transmission with a torque converter gives the loaders fast-movement capability in both forward and reverse, with a minimum of shock. This lets the machines maintain a high production rate. The hydraulic system gives the operator positive control of mounted attachments and assists with steering. Most loaders have pintles or towing hooks for towing small trailers or light loads.

ATTACHMENTS

5-2. The most common loader attachments are a shovel-type bucket or a forklift *(Figure 5-2, page 5-3)*. The loader's hydraulic system provides the power necessary for operating these attachments. Hooks (designed for lifting and moving sling loads) and snowplows are other available attachments.

BUCKET

5-3. Buckets may be *general-purpose* (one-piece, conventional) or *multipurpose* (two-piece, hinged-jaw) *(Figure 5-2)*. The bucket attaches to the tractor unit by lift arms. Buckets are made of heavy-duty, all-welded steel and vary in size from 2.5 to 5 cubic yards. The bucket teeth are bolted or welded onto replaceable cutting edges. Bolt-on, replaceable teeth are provided for excavation of medium-type materials. The multipurpose bucket provides the capability to use the loader as a dozer and to grab material.

FORKLIFT

5-4. A forklift can be attached to the tractor unit in place of a bucket. Designed for material handling, the fork attachment is made of steel with two movable tines.

Cab
(ROPS)

Multipurpose
bucket

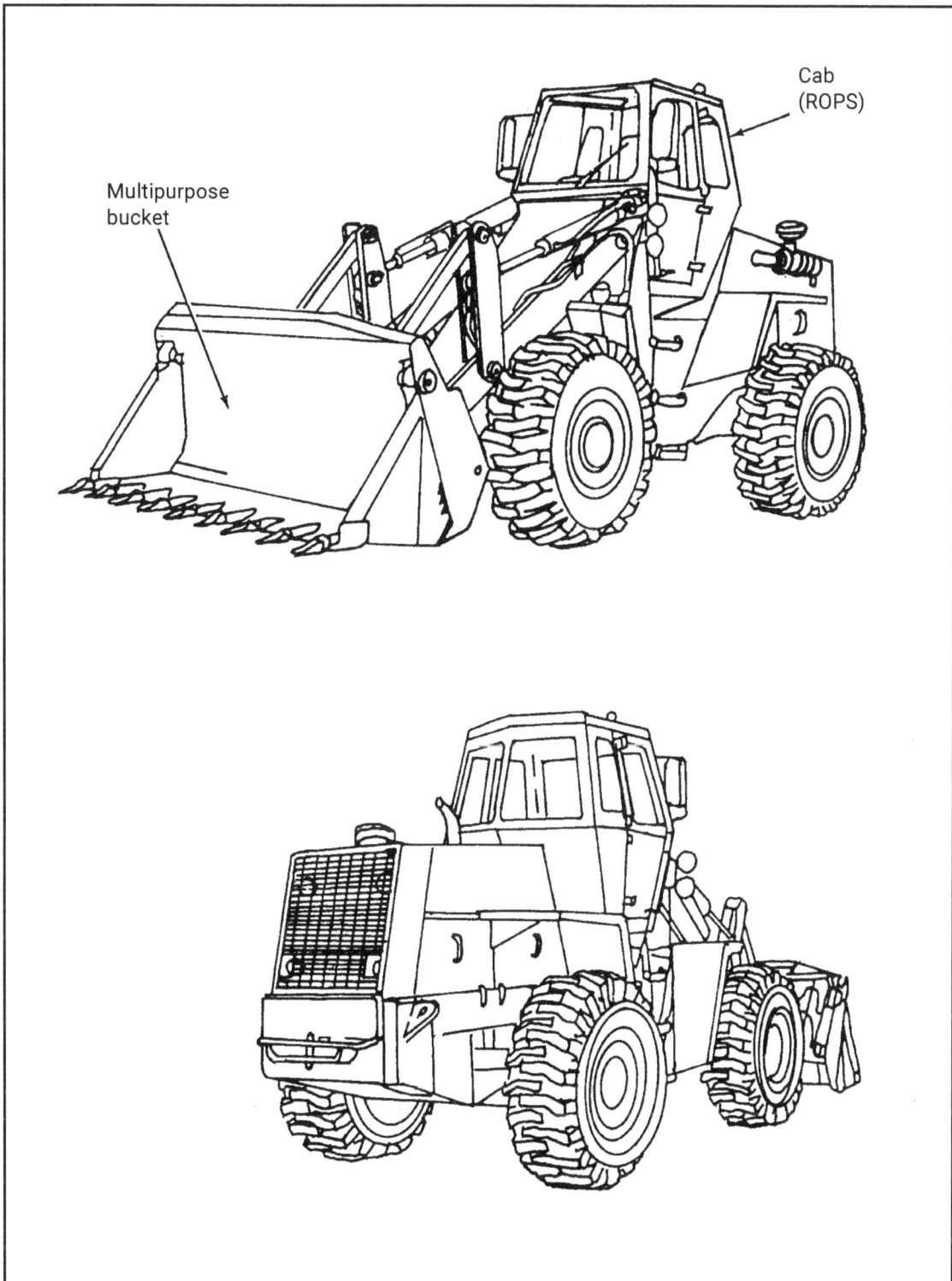

Figure 5-1. Wheel Loader

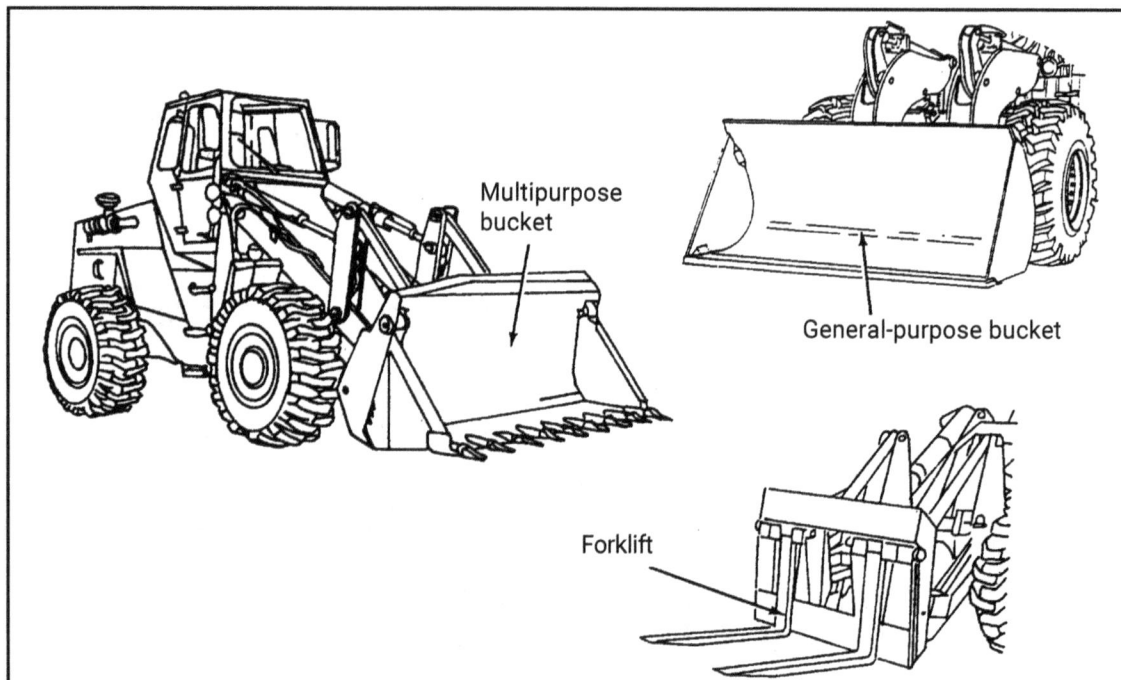

Figure 5-2. Loader Attachments

USE

5-5. Typical uses for a loader are loading trucks; stockpiling materials; digging basements or gun emplacements; backfilling ditches; lifting and moving construction materials; and, when equipped with rock-type-tread tires, operating in and around rock quarries. They may also be used for many miscellaneous construction tasks. These include stripping overburden, charging hoppers and skips, lifting and moving forms for concrete work, moving large concrete and steel pipes, assisting with plant erection and maintenance, and towing small trailers and light loads.

SELECTION

5-6. Two critical factors to consider in selecting a loader are the type and volume of material being handled. Loaders are excellent machines for excavating soft to medium-hard material. Loader production rates decrease rapidly when excavating medium to hard material. Another factor to consider is how high the material must be raised. To be of value in loading trucks, the loader must be able to dump over the side of the truck's dump bed. A loader attains its highest production rate when working on a flat, smooth surface with enough space to maneuver. In poor underfoot conditions or when there is a lack of space to operate efficiently, other equipment may be more effective.

OPERATION

LOADING THE BUCKET

5-7. When loading the bucket, it should be parallel with the ground so its cutting edge can skim the travel surface and remove ruts, obstacles, and loose material on the forward pass. As the cutting edge contacts the bank or stockpile, move the loader forward at a slow speed and increase the power.

While penetrating the material, raise the bucket. Crowd the material into the bucket and roll the bucket back to prevent spilling. Maintain the bucket in an upward position while backing away, to prevent spilling.

POSITIONING OF HAUL UNITS

5-8. Proper positioning of the equipment that will receive material from the loader is necessary for maximum production. This cuts down on maneuver time.

LOADING METHOD

5-9. When loading trucks from a bank or a stockpile with a single loader, use the V-loading method. Use the following steps for the V-loading method *(Figure 5-3)*.

Step 1. With the bucket lowered 1 to 2 inches off the ground, head the loader toward the bank or stockpile in low gear.

Step 2. Move the loader into the stockpile and manipulate the lift and tilt control levers, simultaneously curling back the bucket and raising the boom slightly until the bucket is full and completely rolled back. Maintain power without spinning the tires.

Step 3. Hold the bucket in the upright and curled position, and back away from the stockpile or bank.

Step 4. Approach the haul unit at a 90° angle, lifting the bucket high enough to clear the haul unit.

Step 5. Proceed slowly forward until the bucket is over the haul unit. **Do not** touch the haul unit with the front tires.

Step 6. Dump the bucket by rolling the bucket slowly forward. **Do not** let the bucket hit the haul unit.

Step 7. Back away from the haul unit while simultaneously lowering the boom and leveling the bucket.

Repeat the above steps until the haul unit is loaded.

NOTE: While these machines are flexible and can dig under very awkward conditions, the best production is achieved by keeping both the angle of turn and the travel distance to a minimum.

CLAM LOADING

5-10. This procedure can be used with the multipurpose bucket for handling rocks, timbers, or stockpiles of loose material.

Step 1. Center the front of the bucket on the middle of the first load to be picked up. When about 5 feet from the load, begin to open the bucket.

Step 2. Move the loader forward and make contact with the load. About two-thirds of the opened bucket should penetrate into the material to be loaded.

Step 3. Close the bucket to secure the load.

Step 4. Position the load 10 to 14 inches above the ground.

> ## CAUTION
> Keep the loader bucket as low as possible. A low bucket position provides better balance and operator visibility. When traveling with a full bucket over rough terrain or terrain that can cause the loader to slide, always operate at low speed. Failure to do so can result in loss of control, causing serious injury or loss of life and property damage.

Step 5. Maneuver the loader to the desired location for load placement.

Step 6. Open the bucket fully.

Step 7. Raise the bucket high enough to clear any previously dumped material. Ensure that all of the material is out of the bucket.

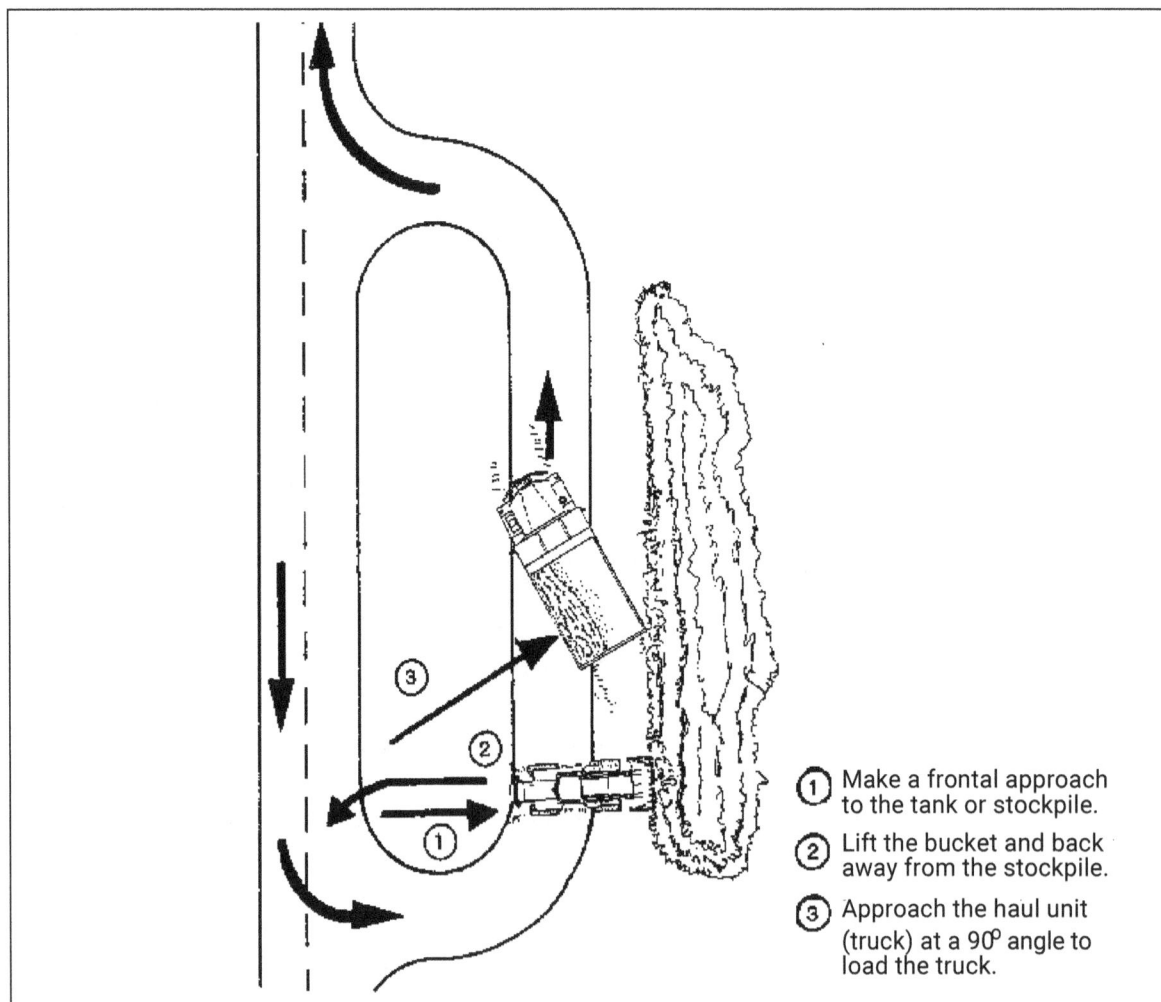

Figure 5-3. Loading Trucks With a Loader (V-Loading Method)

① Make a frontal approach to the tank or stockpile.

② Lift the bucket and back away from the stockpile.

③ Approach the haul unit (truck) at a 90° angle to load the truck.

Step 8. Close the bucket.

Step 9. Place the bucket in the traveling position (10 to 14 inches above the ground).

Repeat the above steps until the task is complete.

EXCAVATING BASEMENTS AND GUN EMPLACEMENTS

5-11. A loader can dig excavations such as basements or gun emplacements if the material is not too hard. The operator should first construct a ramp into the excavation *(Figure 5-4)*. Because the loader works best when excavating above wheel level, the ramp allows the loader to work in that manner and later provides egress to bring out the material. The following procedures are used to construct a ramp.

Step 1. Determine a starting point for the ramp.

Step 2. Position the bucket so it is pitched forward.

Step 3. Move the loader forward, gradually penetrating the earth by lowering the lift control lever.

- Keep the loader in as high a gear as possible without causing the tires to spin excessively.
- Regulate the depth of cut using the lift control lever.

Step 4. Retract the bucket fully.

Step 5. Place the lift control lever in the raised position until the bucket is high enough to clear the surrounding area.

Step 6. Dump the loaded bucket onto a stockpile or into a haul unit.

Repeat the above steps until the excavation is complete.

WORKING IN DIFFICULT MATERIAL

5-12. The multipurpose bucket handles sticky material (which has a tendency to cling to the bucket) better than the general-purpose bucket. A clam-type digging motion works best in this material type. When digging medium to hard material, a greater efficiency can be achieved by first breaking or loosening the material.

Figure 5-4. Constructing a Ramp into an Excavation

BACKFILLING

5-13. When backfilling trenches, lower the bucket to grade level and use the forward movement of the machine to push the stockpiled earth into the trench *(Figure 5-5)*. This type of work is ideal for the loader as long as the bucket is as wide as, or wider than, the loader's wheels or tracks. Narrow buckets cause the wheels to ride up the stockpile. This raises one corner of the bucket and requires more passes. Use the following steps to perform backfilling operations.

Step 1. Align the loader with the stockpile (either to the left or right side) while approaching at a 45° angle so that one-third of the bucket will contact the stockpile.

NOTE: This technique will not work when pushing a large stockpile. In this case, work from the edges.

Step 2. Adjust the bucket by moving the lift control lever to lower the bucket to just off of the natural ground. If using a multipurpose bucket, move the bucket control lever to open the bucket to the clam position.

Step 3. Move the loader forward and gradually move the material. Keep the loader in as high a gear as possible without causing the tires to spin excessively.

Step 4. Move the lift control lever to lower or raise the bucket to cut and spread the material the length of the trench.

Step 5. Move the lift control lever to raise the bucket 10 to 14 inches off the ground before reversing direction.

Step 6. Reverse the loader and return to the stockpile.

Repeat the above steps until the operation is complete.

Figure 5-5. Backfilling a Trench With a Loader

CONSTRUCTING A STOCKPILE

5-14. A stockpile can be constructed from the material excavated in any of the previously described operations. Use the following dump steps when constructing a stockpile:

Step 1. Move the loader forward until the front tires contact the bank.

Step 2. Move the lift control lever to raise the bucket all the way.

Step 3. Move the tilt control lever to slowly tilt the bucket to the dump position.

Step 4. Pull the tilt control lever to tilt the bucket back to the standard bucket position.

Step 5. Back the loader from the stockpile and lower the bucket to about 10 to 14 inches off the ground.

Step 6. Back the loader to the start of the work area.

Repeat the above steps until all of the material is stockpiled.

PRODUCTION ESTIMATES

5-15. Many factors affect loader production: operator skill, extent of prior loosening of the material, slope of the operating area, height of the material, climate, and haul-unit positioning. *Table 5-1* shows bucket fill factors for converting rated heaped-bucket capacity to LCY volume based on the type of material being handled. *Table 5-2* gives average cycle times for wheel loaders to excavate and load with no extra travel required. Use the following formulas and step-by-step method for estimating loader production.

Table 5-1. Bucket Fill Factors for Wheel Loaders

Material	Wheel Loader Fill Factor *
Loose material:	
Mixed moist aggregates	0.95 to 1.00
Uniform aggregates:	
up to 1/8 inch	0.95 to 1.00
1/8 to 3/8 inch	0.90 to 0.95
1/2 to 3/4 inch	0.85 to 0.90
1 inch and over	0.85 to 0.90
Blasted rock:	
Well blasted	0.80 to 0.95
Average	0.75 to 0.90
Poor	0.60 to 0.75
Other:	
Rock-dirt mixtures	1.00 to 1.20
Moist loam	1.00 to 1.10
Soil	0.80 to 1.00
*Decimal of heaped-bucket capacity, for adjustment to LCY	

Table 5-2. Average Cycle Times for Wheel Loaders

Loader Size, Heaped-Bucket Capacity (Cubic Yards)	Wheel-Loader Cycle Time (Minutes)
1.00 to 3.75	0.45 to 0.50
4.00 to 5.50	0.50 to 0.55
NOTE: Includes load, maneuver with four reversals of direction (minimum travel), and dump.	

CUBIC-YARD ESTIMATES

Step 1. Determine the material type and the rated heaped-bucket capacity of the loader.

Step 2. Select the bucket fill factor from *Table 5-1* based on the material type.

Step 3. Determine the average cycle time from *Table 5-2* based on the size of wheel loader.

NOTE: If necessary, add any round-trip travel time to the cycle time found in *Table 5-2* (as would be the case when charging aggregate bins at a plant). Determine the total travel time using the same formula found in *Chapter 3, paragraph 3-28, step 7.* Base the speed on the gear capabilities of the loader matched to the site's travel conditions. For travel distances of less than 100 feet, a wheel loader (with a loaded bucket) should be able to travel at about 80 percent of its maximum speed in low gear and return (with an empty bucket) at about 60 percent of its maximum speed in second gear. Expect slightly higher speeds for longer travel distances.

Step 4. Determine the maximum production rate using the following formula.

$$\text{Maximum production rate (LCY per hour)} = \frac{\text{heaped-bucket capacity} \times \text{bucket fill factor} \times 60 \text{ minutes}}{\text{loader cycle time (minutes)}}$$

Step 5. Determine an efficiency factor. Efficiency depends on both the job conditions and management. A good, average loader efficiency is 50 minutes of work per hour. However, the estimator must always subjectively evaluate all of the conditions which impact efficiency, such as—

- Work-site dimensions, the depth of cut, and the amount of movement required.
- Surface conditions and weather, including the season of the year and drainage.
- Equipment condition.

Step 6. Determine the net production rate in LCY per hour. Multiply the maximum production rate (LCY per hour) by the efficiency factor.

$$\text{Net production rate (LCY per hour)} = \text{maximum production rate (LCY per hour)} \times \text{efficiency factor}$$

EXAMPLE

What is the net production rate (LCY per hour) for a 2.5-cubic-yard wheel loader loading moist loam? Assume average working conditions.

Material type = moist loam
Rated heaped-bucket capacity = 2.5 cubic yards
Bucket fill factor for moist loam *(Table 5-1)* = 1 to 1.1, use an average of 1.05
Cycle time for a 2.5-cubic-yard wheel loader *(Table 5-2)*
= 0.45 to 0.5 minute, use an average of 0.475 minute

$$\text{Maximum production rate (LCY per hour)} = \frac{2.5 \times 1.05 \times 60 \text{ minutes}}{0.475 \text{ minute}} = 332 \text{ LCY per hour}$$

$$\text{Net production rate (LCY per hour)} = 332 \text{ LCY per hour} \times \frac{50 \text{ minutes}}{60 \text{ minutes}} = 277 \text{ LCY per hour}$$

TONNAGE ESTIMATES

5-16. Determine the net production rate in tons per hour. Multiply the net production rate (LCY per hour) by the material weight (tons per LCY). Divide the material weight by 2,000 to convert pounds per LCY to tons per LCY.

$$\text{Material weight (tons per LCY)} = \frac{\text{weight (pounds per LCY)}}{2,000 \text{ pounds per ton}}$$

Net production rate (tons per hour)   net production rate (LCY per hour)   material weight (tons per LCY)

EXAMPLE

What is the net production rate in tons per hour of a wheel loader working in loam (2,200 pounds per LCY) with a net production rate of 263 LCY per hour?

$$\text{Material weight (tons per LCY)} = \frac{2,200 \text{ pounds per LCY}}{2,000 \text{ pounds per ton}} = 1.1 \text{ tons per LCY}$$

Net production rate (tons per hour)   263 LCY per hour   1.1 tons per LCY   = 289 tons per hour

---

OTHER ESTIMATES

- Determine a soil conversion factor if necessary (see *Table 1-1, page 1-4).*
- Determine the total time required. The formula used to determine dozer production *(Chapter 2, paragraph 2-39, step 10)* is applicable.
- Determine the total number of wheel loaders required to complete the mission in a given time. The formula used to determine dozer production *(Chapter 2, paragraph 2-39, step 11)* is applicable.

SAFETY

5-17. Operate wheel loaders carefully because they are easy to overturn. Do not extend any portion of your body between the cab and the lift arms. Operators should adhere to the following safety practices:

- Block the bucket when working on the machine.
- Use caution when removing and replacing the lock rings.
- Do not work between the wheels and the frame while the engine is running.
- Use caution when operating close to the edge of a trench or when working under overhangs created by digging into banks or stockpiles.
- Travel with the bucket at or below axle height.
- Do not carry or lift personnel in the bucket.
- Ground the bucket and set the parking brake before leaving the machine.

Chapter 6

Forklifts

Forklifts are effective over unprepared or unstabilized surfaces. They work well in rough terrain where high-flotation tires are necessary. Most forklifts are four-wheel-steering machines. The Army uses forklifts for loading, unloading, and transporting crates and palletized loads. Examples of such situations would be over a beach; in a surf; and in deep sand, snow, or mud. Without a load, a forklift can move at high speeds between construction sites.

USE

6-1. Forklifts *(Figure 6-1)* were once restricted to use in warehouses or terminals. The Army now uses them for various activities, including—

- Loading and unloading flatcars, trucks, flat trailers, aircraft, and naval landing craft.
- Stocking and transporting heavy crates and palletized loads.

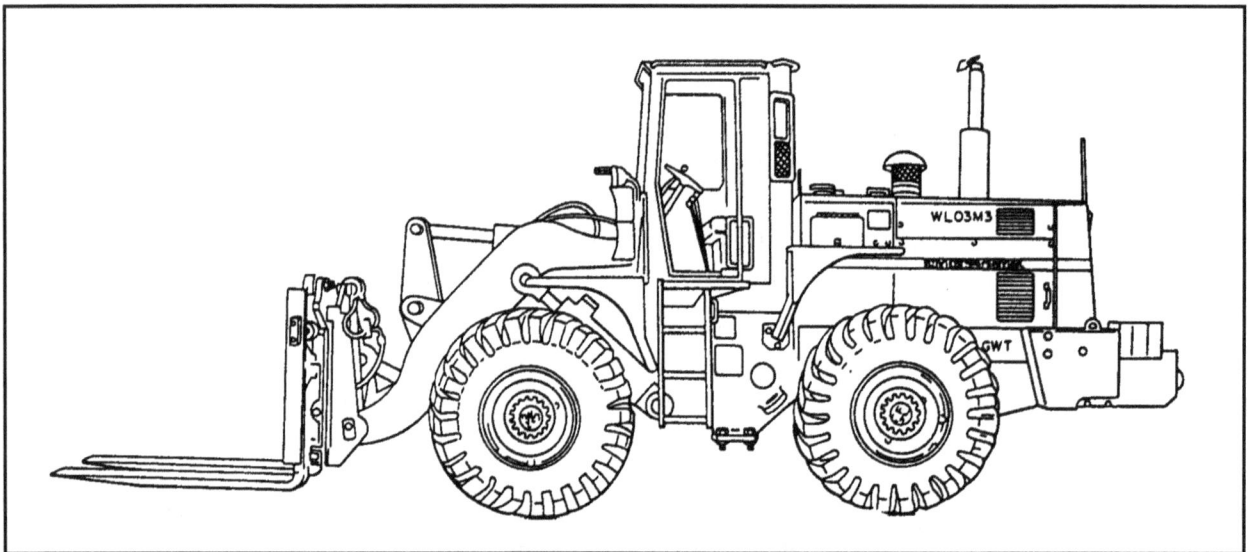

Figure 6-1. Forklift

OPERATION TECHNIQUES

POSITIONING TO LOAD

6-2. When positioning a forklift to pick up a load, bring it in square to the load. Then use the side shift to align the forks rather than trying to align the entire vehicle. Extend the boom as necessary when lifting a load, and retract the boom against the frame when transporting loads. Use oscillation to pick up loads easily at an angle.

TRANSPORTING A LOAD

6-3. When transporting a load, tilt the mast as far back as the load will permit and raise the load only high enough (4 to 6 inches) to clear obstructions. Always change speed gradually, as sudden starts and stops will cause the load to shift. Gradual starts and stops also prevent rapid wear of machine components. Use four-wheel steering for normal material handling and two-wheel steering for high-speed runs.

OPERATING IN SAND OR MUD

6-4. Lower the tire pressure when operating in sand or mud. Check the operator's manual for appropriate pressures.

TRANSPORTING ON RAMPS AND GRADES

6-5. When using a forklift to transport cargo up ramps or other grades, carry the load on the upgrade end of the machine. When carrying cargo downgrade, back the forklift down the grade with the load on the upgrade end. Carry all loads with the tines tipped back.

OPERATING IN WATER

6-6. For operating in water, disconnect the fan and use four-wheel drive. (Check the operator's manual for servicing requirements after operating in salt water.)

SAFETY

6-7. Operators must always face in the direction of travel. Carry the load so that it does not obstruct the operator's vision in the direction of travel. When forklifts are not in operation, lower the forks and rest them flat on the ground or floor.

OVERHEAD SAFETY GUARDS

6-8. Equip forklifts of all types with steel overhead safety guards. Permit exceptions only when the overhead safety guards either increase the overall height of the forklift or restrict the operator's freedom of movement.

LOAD CAPACITY

6-9. Stencil the machine's load capacity and gross weight on the machine in plain view of the operator. Never exceed this capacity. Do not counterweight the machine to increase lifting capacity. The capacity rating is based on the load positioned 24 inches from the fork's heel.

HOISTING PERSONNEL

6-10. Use a forklift to hoist personnel only after obtaining a supervisor's approval and under the following conditions:

- Allow only skilled personnel to perform tasks requiring elevation of personnel by forklift.
- Use special *personnel pallets* (guardrails on all four sides).
- Face all personnel away from the mast with their hands clear of the hoisting mechanism during the actual raising and lowering.

Chapter 7

Cranes

Cranes are used to hoist and place loads. They are mounted in one of three ways—truck, crawler, or rough-terrain (wheel). The truck and rough-terrain mounts do not provide the stability of the crawler mount. Attaching accessory equipment to the crane's superstructure and boom allows use of the basic machine for pile driving or as an excavator.

BASIC CRANE UNIT

7-1. The basic crane unit *(Figure 7-1)* consists of a substructure mount and a full revolving superstructure. The upper revolving superstructure is substantially the same without regard to the substructure mount. Installing attachments allows use of the machine for tasks other than hoisting. *Figure 7-2, page 7-2,* shows several crane attachments —hook block, clamshell, pile driver, and dragline.

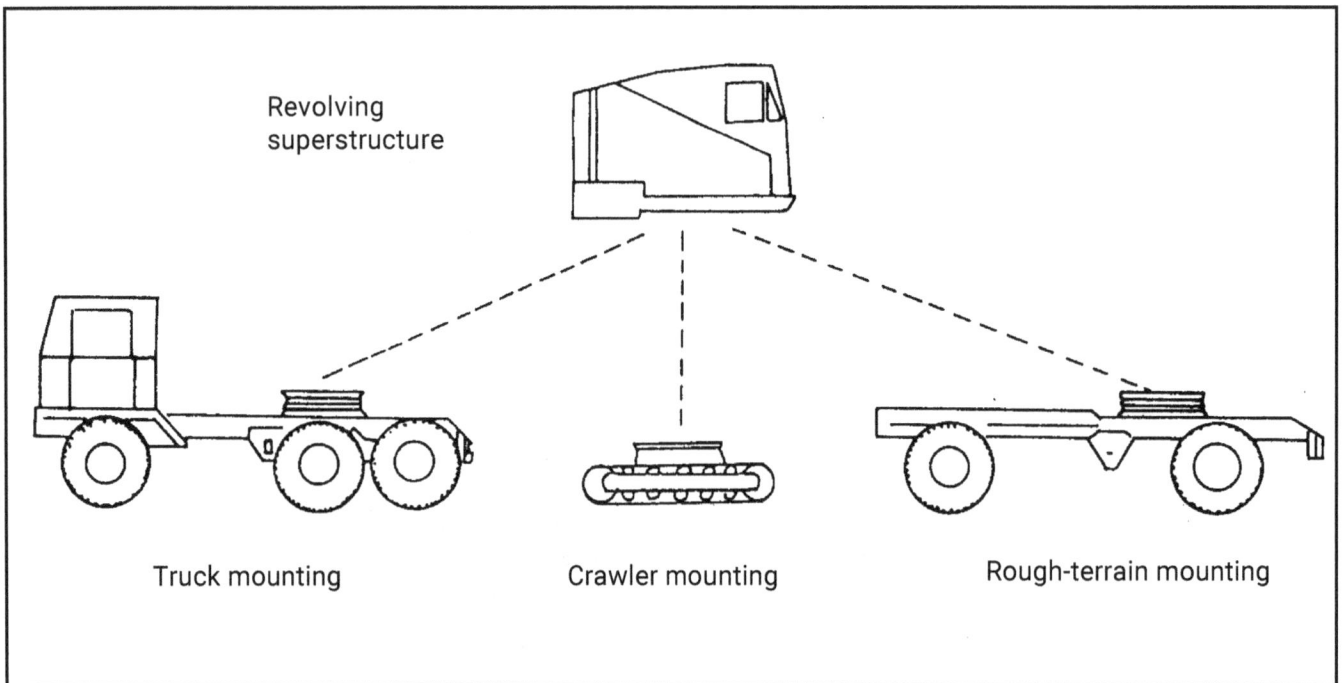

Figure 7-1. Basic Crane Unit

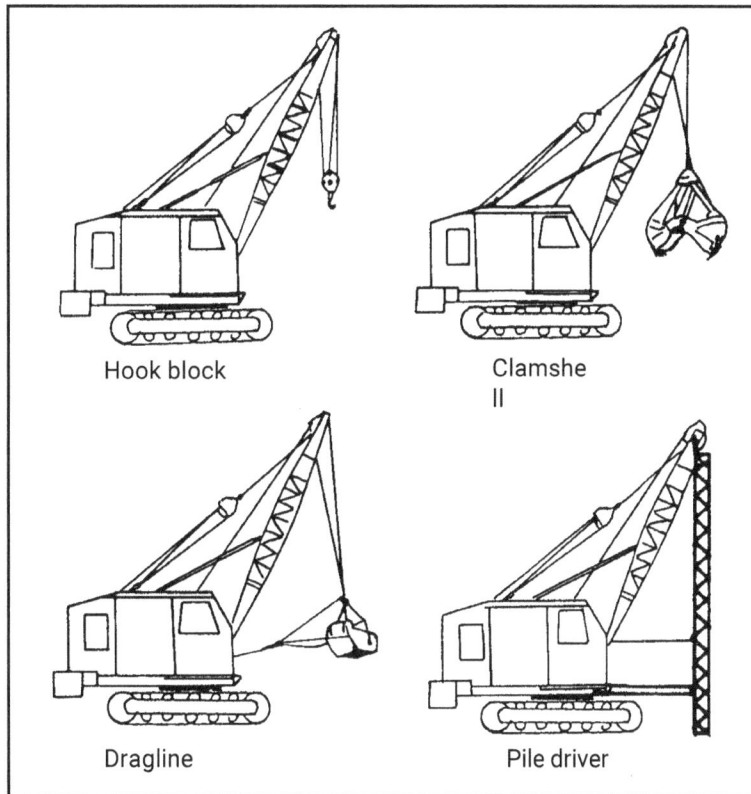

Figure 7-2. Crane Attachments

SUBSTRUCTURE

7-2. The substructure can be a truck, a crawler, or a rough-terrain mount. Each mount has different stabilization and terrain capabilities.

Truck Mount

7-3. Truck cranes have specially-designed heavy-duty truck mounts. This mounting provides good between-project mobility. The 20- and 25-ton truck cranes can operate a hook block, a 0.75-cubic-yard clamshell, a 7,000-foot-pounds-per-blow diesel pile driver, or a dragline. *Figure 7-3* shows a 25-ton truck crane with a hydraulic telescopic boom.

- **Stability.** These cranes have outriggers on each side. Always employ the outriggers when operating the crane or attachments. Fully extend the outriggers and secure the jacks on the base plates so that the crane is completely off of the tires. Some models are equipped with hydraulic outriggers.
- **Terrain capability.** Because of its limited stability, the truck mount restricts the efficient movement of these cranes to firm, level terrain.

7-4. **Towing.** A pintle hook (towing connection) is on the rear of the truck, and towing eyes are on the front. The pintle hook enables the truck to tow an attachment trailer for transporting associated attachments to the job site. If the truck becomes inoperable or stuck, use the towing eyes to attach the truck to a towing vehicle. The towing eyes will withstand twice the dead-weight pull of the vehicle.

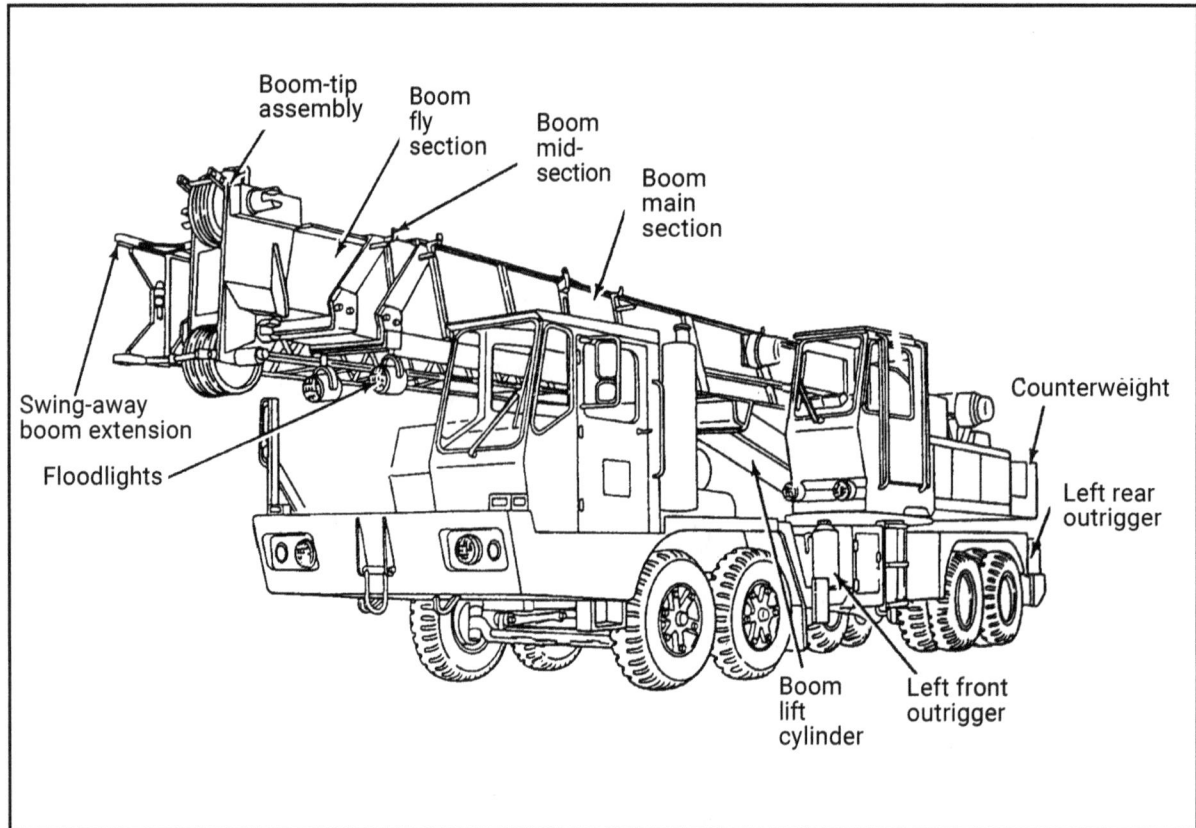

Figure 7-3. Truck Crane (25-ton) With a Hydraulic Telescopic Boom

7-5. Operating Hints.

- Be sure the crane is level prior to operating.
- Do not hoist loads over the front. Generally, perform all heavy hoisting over the rear of the truck since the truck's cab and engine provide additional counterweight for the load. When using a load chart, ensure that the capacity reflects the quadrant of the proposed hoist. For example, over the side, over the rear of the mount, or over the front of the mount.
- Use the outriggers.
- Check the base plates periodically to ensure that good soil bearing is being maintained.
- Place the truck's transmission in neutral when operating the superstructure. This prevents possible damage to the gears by any rocking movement of the truck.

Crawler Mount

7-6. Move crawler cranes from project to project on transport trucks. Crawler cranes have relatively low travel speeds, and long travel causes excessive track wear. Crawler cranes are best suited for longer duration jobs. The crawler mount provides excellent maneuverability on the job site and has low ground-bearing pressure. The 40-ton crawler crane can operate a hook block, a

2-cubic-yard clamshell, a 24,000-foot-pounds-per-blow diesel pile driver, or a dragline. The military normally uses the 40-ton crane in quarry operations.

- **Stability.** The crawler mounting provides a stable base for operating the revolving superstructure. Because of the width and length of the crawler tracks, the weight of the machine spreads over a large area.
- **Terrain capability.** Crawler cranes are designed to operate on level ground. A 3° out-of-level condition can result in a 30 percent loss in hoisting capacity. Crawler cranes' low ground-bearing pressure enables them to travel over soft ground. In locations where the ground is extremely soft or unstable, use timber mats to provide a firm footing *(Figure 7-4)*. When not handling a load, the machine can climb slight grades.

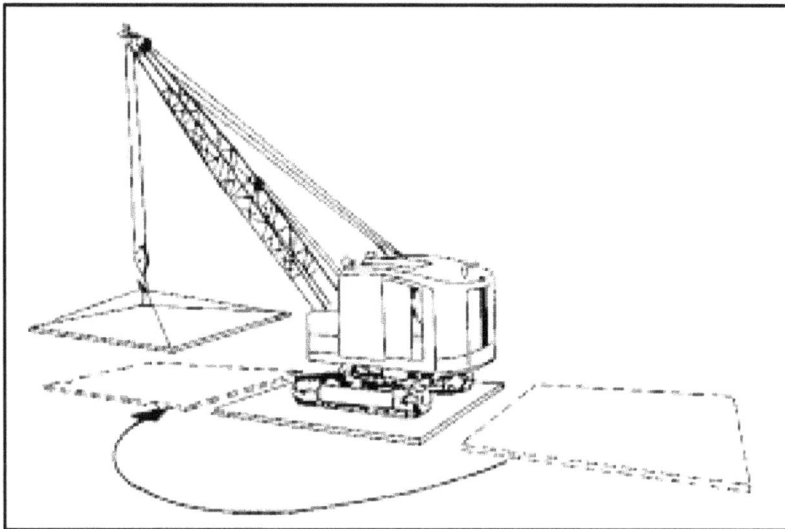

Figure 7-4. Timber Mats Supporting a Crawler Crane on Soft Material

7-7. Crawler cranes can be lashed on barges to operate over water. They can operate in shallow water as long as the water does not enter the revolving superstructure. Before moving the machine into the water, check the water depth and the under-foot conditions. Thoroughly clean and service the machine after working in salt water.

Rough-Terrain Mount

7-8. The two models of rough-terrain cranes are the 22- and the 7.5-ton (Type I, general purpose, and Type II, airborne/airmobile). The 22-ton rough-terrain crane can operate a hook block, a 0.75-cubic-yard clamshell, a 7,000-foot-pounds-per-blow diesel pile driver, or a dragline. The main differences between the rough-terrain crane and the truck crane are the large tires and the high ground clearance, which enables the rough-terrain crane to work in areas inaccessible to the truck crane.

7-9. **Rough-Terrain Crane (22-Ton).** The rough terrain crane has the capability of long-distance highway travel and good job-site maneuverability. The mount provides four-wheel-drive capability, conventional two-wheel steering, four-wheel steering, and crab steering. The travel speed of the machine is 55 mph on the highway. This crane has dual cabs, a lower cab for highway travel, and a superstructure cab that has both the drive and the crane controls. A pintle hook on the rear of the mount lets it tow an attachment trailer. However, towing the trailer cross-country is not recommended.

- **Stability.** Since the mount is not suspended on springs, the front axle must oscillate. This oscillation prevents the mount from tipping or rolling over if one of the front wheels drops into a hole. Use the outriggers to raise the machine off the ground and to level it before extending the boom. If using outriggers to stabilize the machine, position safety wedges on the front axle to prevent oscillation. The outriggers operate individually, allowing the superstructure to be leveled. Load ratings are based on the assumption that the crane is in a level position for the full 360° of swing.
- **Terrain capability.** The machine's low ground-bearing pressure lets it travel over relatively soft terrain. It can traverse slopes up to 48 percent if the ground is firm and dry.

7-10. **Rough-Terrain Crane (7.5-Ton).** The basic 7.5-ton rough-terrain crane (Type I) has a diesel engine, pneumatic tires, two- and four-wheel-drive capability, and two- and four-wheel-steering capability. The cab is located on the mount instead of on the full 360° rotating superstructure. It has a hydraulic boom that is extendible from 28 feet to a maximum of 35 feet. This crane can perform lifting and carrying operations (desirable with standard North Atlantic Treaty Organization [NATO] pallets). The rough-terrain, 7.5-ton airborne crane (Type II) *(Figure 7-5)* is a modification of the basic 7.5-ton unit, making it suitable for airborne and airmobile operations.

- **Stability.** The rough-terrain mount has four independently operated outriggers (two on each side).
- **Terrain capability.** This crane can safely traverse typical construction terrain, longitudinal grades of 30 to 50 percent, and side slopes of 15 to 30 percent in four-wheel drive. This crane can be used on firm terrain because of its high ground-bearing pressure tires.

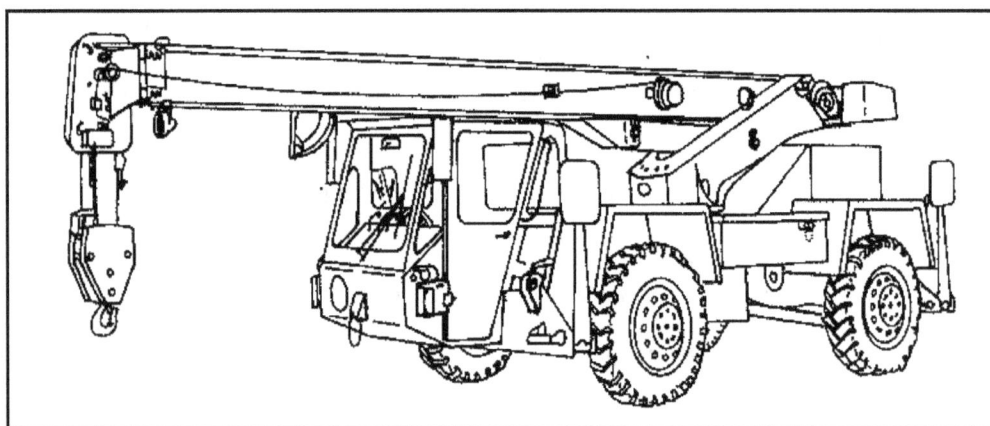

Figure 7-5. Rough-Terrain, 7.5-Ton Airborne Crane

SUPERSTRUCTURE

7-11. The revolving superstructure rests on the mount and includes the counterweight, the engine, the operating mechanism, the boom, the cab, and sometimes a separate engine.

Counterweight

7-12. The counterweight is normally a cast-steel member attached to the rear of the superstructure to produce a countermoment to the weight and radius of the load. This countermoment prevents the crane from tipping.

Operating Mechanisms

7-13. Two independent cable drums control the operation of the various attachments. The drums are mounted parallel to each other or one behind the other. Refer to them by their relative mounting (*right* or *left*, *front* or *rear*) or by their function (*drag cable drum* during dragline operations or *closing-line cable drum* during clamshell operations). When using the drums in conjunction with a hook block, refer to them as the *rear* or *main hoist drum*.

7-14. A third cable drum, the *boom hoist drum*, controls the raising and lowering of the boom for those cranes having a lattice boom. Some models have a two-piece, grooved lagging for quick attachment to the drum shaft. The diameters of the grooved lagging differ depending on the make and model of the machine. Differences in lagging diameter provide different operating line speeds. For example, the lagging used for dragline operations may be smaller to provide a slower line that gives greater power.

7-15. Clutches and brakes may be powered mechanically, hydraulically, or pneumatically. Some makes and models have an internally mounted clutch that, when actuated, expands to engage the drum. Other makes and models have external contracting clutches.

Boom

7-16. **Lattice Boom.** The lattice boom is a latticed structure consisting of four main chords connected with lacing. The basic boom consists of a base section supported on the revolving superstructure and an upper section with a boom head. The sections are fastened together by one of two methods—bolted butt-plate (flange) connections or pin and clevis connections. The length of a lattice boom can be increased in one of two ways—inserting an intermediate section between the upper and base sections (the most common way) or adding a boom tip extension called a jib. A jib is a lighter structural section. An offset jib permits greater load radius than an equivalent length of standard boom. Crane booms not equipped with jib-boom anchor plates can only use intermediate sections for extending the boom. When lengthening lattice booms, extend the gantry or A-frame to provide the required lifting angle for the boom lines.

7-17. **Hydraulic Telescopic Boom.** This boom consists of two or more telescoping boxes made of steel plates. The action of hydraulic cylinders extends or retracts the boxes. Only certain attachments can be used with these booms. This boom type is used on some rough-terrain cranes and truck cranes *(Figure 7-3, page 7-3).*

HOISTING OPERATIONS

7-18. The hook block *(Figure 7-2, page 7-2)* used for hoisting operations is applicable on either a lattice or a telescoping boom. Basic crane equipment includes hoist drums, hook blocks to provide the required parts of line (reeving), and boom-suspension and hoist cables.

FACTORS AFFECTING HOISTING CAPACITY

7-19. Boom length, operating radius or boom angle, type of mount, stability (use of outriggers), amount of counterweight, hook-block size, hoisting position, and maintenance determine the crane's safe hoisting capacity.

Boom Length

7-20. Increased boom length reduces a crane's hoisting capacity. Use of a jib attachment will further reduce the hoisting capacity. The increased load moment at the greater operating radius and the added weight of the additional boom sections decrease the hoisting capacity.

Operating Radius

7-21. Operating (working) radius *(Figure 7-6)* is the horizontal distance measured from the axis of rotation of the superstructure to a vertical line extending down from the outside edge of the crane's boom-tip (head) sheave. Cranes are rated according to hoisting capacities at various radii, based on either tipping or structural constraints. Swinging the boom causes a centrifugal force *(Figure 7-7, page 7-8),* which in effect increases the operating radius. High winds can also push the load, which increases the operating radius *(Table 7-1, page 7-8).* As the working radius increases, the hoisting capacity decreases. It is essential to know the weight of a load before hoisting.

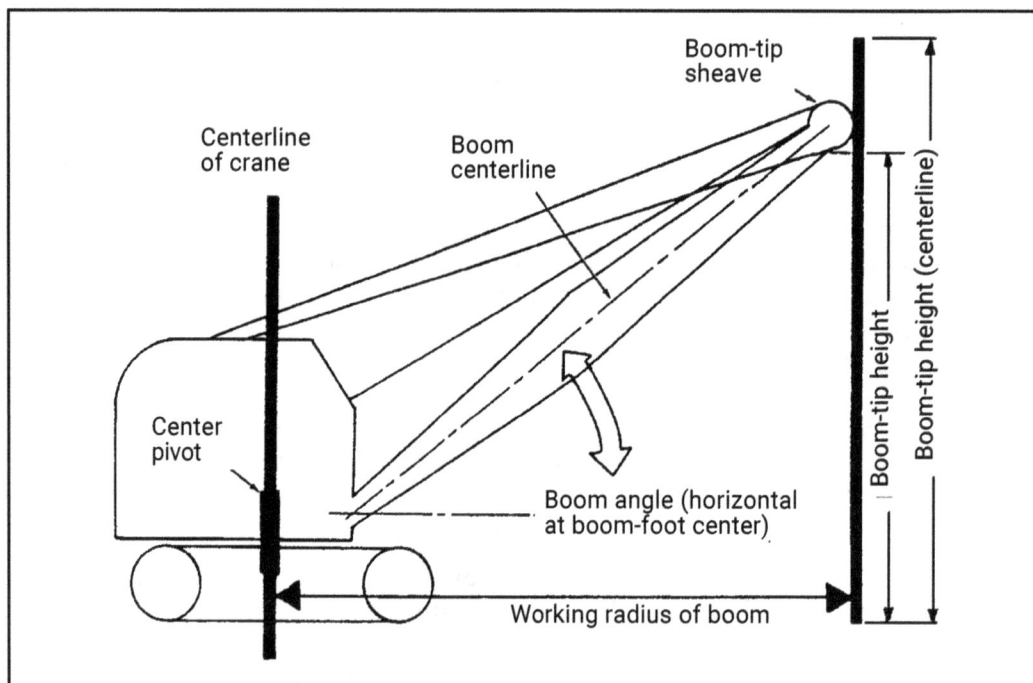

Figure 7-6. Operating Radius of a Boom

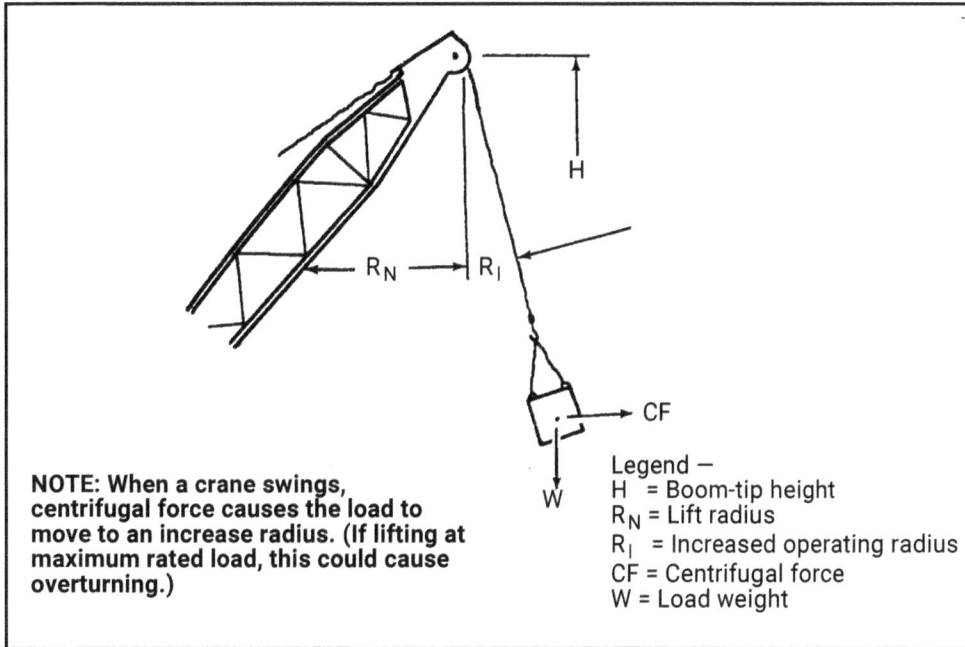

NOTE: When a crane swings, centrifugal force causes the load to move to an increase radius. (If lifting at maximum rated load, this could cause overturning.)

Legend —
H = Boom-tip height
R_N = Lift radius
R_I = Increased operating radius
CF = Centrifugal force
W = Load weight

Figure 7-7. Effects of Centrifugal Force on the Operating Radius

Table 7-1. Effects of Wind on Crane Operations

Approximate Wind Speed		Weather Conditions	Effects
mph	kph		
0-1	0-1.5	Calm	Smoke rises straight up
1-3	1.5-5	Light air	Smoke drifts
4-7	6-11	Slight breeze	Leaves rustle
8-12	12-19	Gentle breeze	Leaves and small twigs move
13-18	20-29	Moderate breeze	Dust and papers fly; small branches move
19-24	30-39	Fresh breeze	Small trees sway
25-31	40-50	Strong breeze	Large branches move
32-38	51-61	High wind	Walking is difficult; tree trunks bend
39-46	62-74	Gale	Twigs break off
47-54	75-87	Strong gale	Shingles are blown away
55-63	88-100	Whole gale	Trees may be uprooted

Mount Type

7-22. The basis of a crane's hoisting-capacity rating concerning tipping is as follows (check the operator's manual for the correct table):

- **Crawler mounted.** Power Crane and Shovel Association (PCSA) standards limit the load capacity concerning tipping to 75 percent of the maximum load.

- **Truck and rough-terrain mounted.** PCSA standards limit the load capacity concerning tipping to 85 percent of the maximum load whether on tires or outriggers.

Capacity at short radii is usually limited by the structural capacity of the machine's components. In the case of hydraulic cranes, the hydraulic capacity sometimes limits the lifting capability.

Stability

7-23. It is extremely important that the crane be positioned on a firm, level footing and, if outriggers are available, that they be fully extended and bearing. If necessary, prepare the crane's operating site in advance. Adherence to a machine's load-chart ratings ensures that the crane is stable in terms of a moment balance between load weight and radius and the machine's counterweight. A crane's hoisting ability as presented in the load chart assumes that the machine is positioned on solid, level ground. Take care not to position the crane's tracks or outriggers over underground utilities or recompacted trench excavations or close to excavation edges, all of which could give way when hoisting a load.

Counterweight

7-24. Do not add additional counterweight or anchor the crane to a *deadman* in an attempt to increase hoisting capacity. These procedures could cause structural damage to the crane.

Hook-Block Size

7-25. The hook block on the crane should be the size prescribed for the crane. Lack of proper rigging and hook capacity may damage the block and the sheave system. Tables of hoisting capacities are based on gross capacity; therefore, the weight of the hook block and slings must be considered as part of the load.

Hoist Position

7-26. The load capacity of a crane depends on the quadrant position of the boom with respect to the machine's undercarriage. In the case of a crawler crane, consider three quadrants; over the side, over the drive end of the tracks, and over the idler end of the tracks. Usually, imaginary lines from the superstructure's center of rotation through the position of the outriggers define the quadrants for mobile cranes. Always consider the minimum condition based on swinging the load from point of pick up to final placement.

Maintenance

7-27. Proper crane maintenance will help to achieve rated, safe hoisting capacity. Check the following items to ensure that the crane can attain its rated hoisting capacity:

- Type, size, and condition of the wire ropes.
- Type, size, and condition of the hook block.
- Structural condition of the boom.
- Mechanical condition of the engine.
- Adjustments and functional qualities of the clutches and brakes.

SAFE HOISTING-CAPACITY CALCULATION

Step 1. Determine the required clearance *(Figure 7-8)*, which includes the—
- Load height.
- Hoisting height.
- Hook-block height.
- Minimum, estimated safe distance between the boom-tip sheave and the hook-block sheave. (This is a judgment call; normally, a 2-foot clearance is satisfactory.)
- Height of the slings from the load's top to the hook.

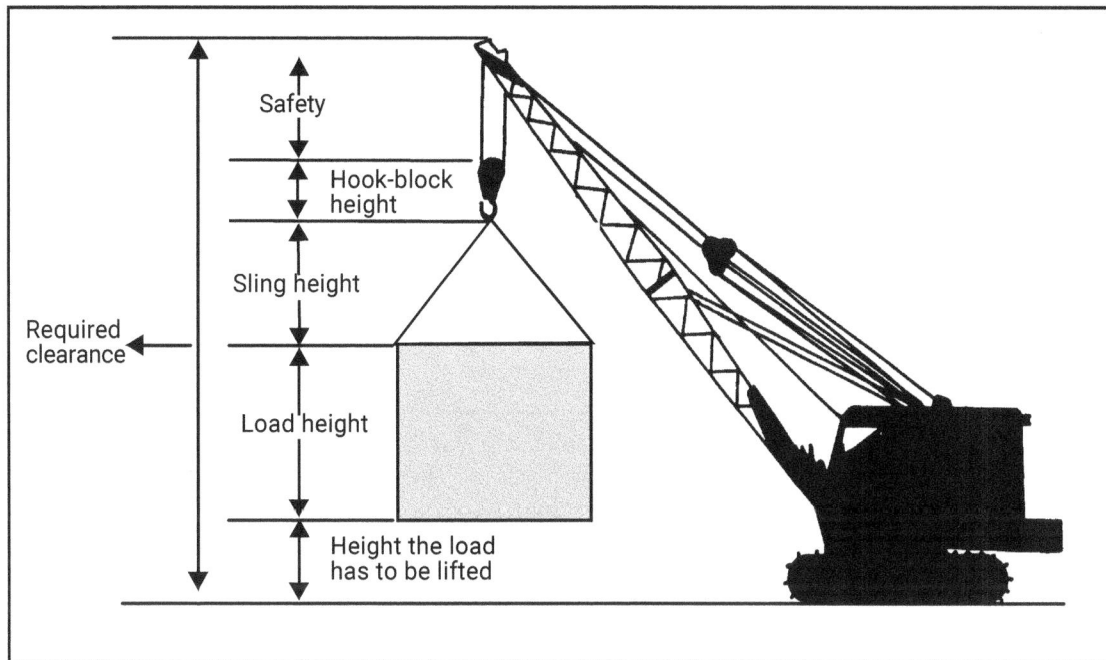

Figure 7-8. Required Clearance

Step 2. Determine the total weight to hoist, which includes the—
- Load weight.
- Hook-block weight.
- Sling weight.
- Stowed jib weight (if mounted on the boom).

Step 3. Determine the working radius. This is the distance from the crane's center of rotation to the load's center. Always consider the maximum radius that results when swinging the load from point of pick up to final placement.

NOTE: Remember that centrifugal force (caused by swinging the boom or from the wind) can increase the radius. Refer to *Figure 7-7, page 7-8,* and *Table 7-1, page 7-8,* respectively.

Step 4. Determine if the crane will hoist the load. Use the appropriate equip-
ment-hoisting charts and the following information:

- Boom length.
- Boom angle.
- Boom-tip height.
- Total hoisted weight.

OPERATION TIPS

- Try to position the crane to eliminate swinging over workers.
- Ensure that the supporting ground has adequate strength.
- Position the crane for the shortest possible boom swing and swing the load slowly when performing repetitive hoisting.
- Ensure that the machine is always level.
- Use tag lines on loads to prevent excessive swaying of the load.
- Use adequate hoist-line lengths to ensure full travel of the block to the lowest point required.
- Organize the work for minimum travel time. When possible, complete all needed hoists in one area before moving to a new position.
- Use the power-down device on the equipment (if available) when performing precise load handling.
- Do not use excessive counterweight or tie-down devices to increase stability.
- Check weather reports or use the indicators given in *Table 7-1, page 7-8,* to determine approximate wind speed. Wind can affect crane operations and even cause overturning when hoisting close to or at maximum hoisting capacity. Cease work if wind speed exceeds 30 mph.

PILE DRIVER

DESCRIPTION

7-28. The pile driver attachment *(see Figure 7-2, page 7-2)* consists of adapter plates, leads, a catwalk, a hammer, and a pile cap. The adapter plates are bolted to the top section of the leads and fastened to the boom tip. The leads are fastened below the base of the boom. Pile leads cannot be attached to a jib boom. The hammer may be diesel or drop type.

Diesel Hammers

7-29. Diesel-driven, pile-driving hammers come in two types—open-top or closed-end. Both types have self-contained, free-piston engines, operating on a two-cycle compression-ignition principle. Diesel hammers eliminate the need for air compressors or steam boilers to power the hammer. Do not use these hammers to pull piles. They are suitable for use on either lattice or telescopic booms. When driving a pile in soft soil, a diesel hammer may not fire because there is insufficient soil resistance to support fuel ignition. When this happens, revert to a drop hammer until the pile reaches sufficient driving resistance.

Drop Hammers

7-30. Gravity-operated drop hammers are best for driving vertical piling. When piling is angled, part of the driving force is lost in friction with the pile leads. Drop hammers are relatively slow compared to other types of hammers. Use a hammer that is at least as heavy as the pile being driven; for best results, the hammer should be twice as heavy as the pile. It is better to use a heavier hammer with a smaller drop. Raise and drop the hammer at a steady rate of speed. Typical drop rates are four to eight blows per minute. The recommended drop height varies with the type of pile; for example, the recommended drop height is 15 feet for timber piles and 8 feet for concrete piles.

DRIVING RATE

7-31. Driving time varies greatly depending on the terrain, the weather, the soil conditions, the type of pile, and the type of hammer used. The only way to determine the driving rate is to drive a pile under project conditions. A rule of thumb to use for planning is 30 minutes to drive a 12-inch diameter pile 20 feet. This includes the time for setting the pile in the leads.

OPERATION TIPS

7-32. Use the following guidelines when operating a pile driver:

- Position the crane so that it will require the minimum time to move between pile locations. Placement is generally parallel to the long axis of the pile group.
- Place the piles close to the driving locations so that the crane only needs to swing to pick up the next pile.
- Make shallow, continuous blows with the hammer. High, infrequent blows cause pile failures.

CLAMSHELL

DESCRIPTION

7-33. The clamshell *(see Figure 7-2, page 7-2)* consists of a clamshell bucket, hoist drum laggings, a tag line, and wire ropes for the boom. The clamshell bucket consists of two scoops hinged together. A clamshell cannot be operated off of a jib. Clamshell drum laggings may be the same as those used for the crane, or they may be changed to meet the speed and pull requirements of the clamshell. This requirement may change with the design of the equipment (check the operator's manual). Usually, the same wire ropes used for hoisting operations can be used for clamshell operations. However, two additional lines must be added—a secondary hoist line and a tag line. The tag line is a small-diameter cable with a spring-tension winder that is used to prevent the clamshell bucket from twisting during operation. The tag line and winder, like the clamshell bucket, are interchangeable with any make or model in the same size range. The spring-loaded tag line does not require operator control and does not attach to the crane's operating drums. The winder is usually mounted on the lower part of the boom.

7-34. The clamshell is a vertically-operated attachment capable of working at, above, and below ground level. Attach the clamshell bucket to the crane's hoist line. A clamshell can dig in loose to medium-stiff soils. The length of the boom determines the height a clamshell can reach. The length of wire rope the cable drums can accommodate limits the depth a clamshell can reach. A clamshell's hoisting capacity varies greatly. Factors such as the boom length, the operating radius, the size of the clam bucket, and the unit weight of the material excavated determine a clamshell's safe hoisting capacity. Refer to the crane's load-capacity table in the operator's manual.

USE

7-35. The clamshell is best for jobs such as excavating vertical shafts or footings or for charging aggregate bins or hoppers. The holding, closing, and tag lines control the bucket movement. At the start of the digging cycle, the bucket is dropped on the material to be dug with the shells open. As the closing line is reeved in, the shells are drawn together causing them to dig into the material. The weight of the bucket, which is the only crowding action available, helps the bucket penetrate the material. The holding and closing lines then raise the bucket. Release the tension on the closing line to open the bucket and dump the material.

Excavating Vertical Shafts or Footings

7-36. Since the dimensions of this type of excavation may vary, it is difficult to give the most efficient digging position for the clamshell. Two important facts to consider are the amount of wire rope on the machine and the need to keep the outside edge of the cut lower than the center (this keeps the bucket from drifting toward the center and causing a V-shaped excavation). In deep excavations, a bucket spotter or signalman usually guides the operator, especially when the bucket is out of the operator's sight. Spotters may also need to use hand tag lines to guide the bucket.

Charging Aggregate Bins or Hoppers

7-37. Position the crane to avoid having to raise and lower the boom when swinging between the aggregate stockpiles and the bins or hoppers.

PRODUCTION ESTIMATES

7-38. The following factors make it difficult to arrive at dependable clamshell production rates:

- The difficulty of loading the bucket in different soil types.
- The height to hoist the load.
- The slow swing required.
- The method of disposing of the load.

For example, when loading material into a truck, the time required to spot the bucket over the truck and dump the load is greater than when dumping the material onto a spoil pile. The best method for estimating production is to observe the equipment on the job and measure the cycle time. Use the formulas in steps 1 through 5 shown in the following example when cycle-time data is available.

EXAMPLE

Determine the number of hours it will take to load 450 LCY of aggregate from a stockpile into haul units with a clamshell.

Bucket size = 0.75 cubic yard
Average cycle time = 40 seconds
Efficiency factor = 50-minute working hour

Step 1. Determine the bucket size.

Bucket size = 0.75 cubic yard

Step 2. Determine the working time (in seconds per hour). Convert working minutes per hour to working seconds per hour.

Working time (seconds per hour)
= working minutes per hour ×60 seconds per minute
= 50 working minutes per hour ×60 seconds per minute
= 3,000 seconds per hour

Step 3. Determine the production rate.

Production rate (LCY per hour)

$$= \frac{B \times T \text{ (seconds per hour)}}{CT \text{ (seconds)}}$$

$$= \frac{0.75 \text{ cubic yard} \times 3{,}000 \text{ seconds per hour}}{40\text{-second cycle}}$$

$$= 56 \text{ LCY per hour}$$

where—
B = bucket size
T = working time
CT = cycle time

Step 4. Determine the soil conversion factor, if needed.

Soil conversion factor = not applicable

Step 5. Determine the total time required to complete the job.

$$\text{Total time (hours)} = \frac{\text{quantity of material moved (LCY)}}{\text{production rate (LCY per hour)}} = \frac{450 \text{ LCY}}{56 \text{ LCY per hour}} \quad 8 \text{ hours} \quad =$$

OPERATION TIPS

- Position the unit on level ground.
- Position the unit so digging operations are at the same radius as the dumping operation. This will avoid wasted production caused by raising and lowering the boom.
- Select the correct bucket size for the machine. Efficient use of the clamshell means an efficient digging, hoisting, swinging, and dumping cycle. Large buckets may increase the cycle time.
- Remove the bucket teeth when working in soft materials.

DRAGLINE

7-39. Dragline components consist of a drag bucket and a fairlead assembly. Wire ropes are used for the drag, the bucket hoist, and the dump lines. The fairlead guides the drag cable onto the drum when the bucket is being loaded. The hoist line, which operates over the boom-point sheave, raises and lowers the bucket. In the digging operation, the drag cable pulls the bucket through the material. When the bucket is raised and moved to the dump point, it is emptied by releasing the tension on the drag cable. Dragline buckets are rated by type and class, as follows:

- **Bucket types.** Type I (light duty), Type II (medium duty), and Type III (heavy duty). The Army usually uses Type II buckets.
- **Bucket classes.** Class P (perforated plate) and Class S (solid plate). The Army usually uses Class S buckets.

USE

7-40. The dragline (see *Figure 7-2, page 7-2*) is a versatile attachment capable of a wide range of operations at or below ground level. It can handle material ranging from soft to medium-hard. The greatest advantage of a dragline over other machines is its long reach for digging and dumping. A dragline does not have the positive digging force of a shovel or backhoe. Breakout force is derived strictly from bucket weight. The bucket can bounce, tip over, or drift sideward when it encounters hard material. These weaknesses are particularly noticeable with lightweight buckets.

7-41. Use a dragline for trenching, stripping overburden, cleaning and digging roadside ditches, and sloping embankments. The dragline is the most practical attachment to use when handling mud. The dragline's reach allows it to excavate an extensive area from one position. The sliding action of the bucket decreases suction problems.

NOTE: Do not use the dragline attachment with hydraulic cranes.

CAPACITY

7-42. The dragline boom may be angled relatively low when operating. Boom angles of less than 35° from the horizontal plane are seldom advisable because of the possibility of tipping the machine. When excavating wet, sticky material and casting it onto a spoil bank, the chance of tipping increases because of material sticking in the bucket.

Casting Material

7-43. The throw or cast of the bucket increases the dragline's operating radius, which can be up to one-half of the boom height *(Figure 7-9, page 7-16)*.

> **WARNING**
> Only experienced operators should perform extended casting because of the possible damage to the cables or the boom or the possibility of tipping the machine.

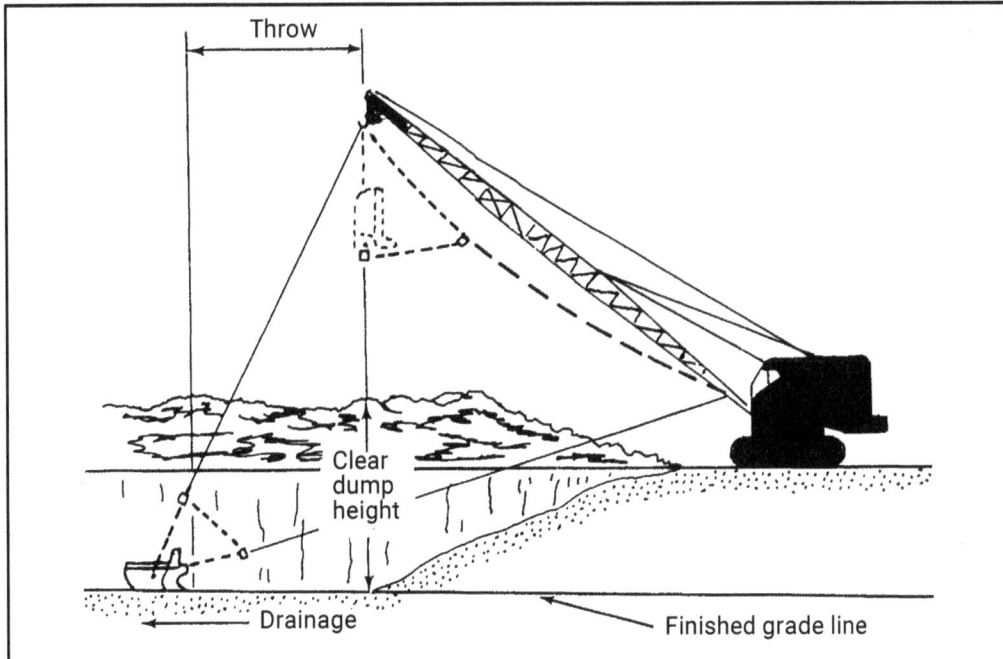

Figure 7-9. Dragline Throw

Excavating a Trench

7-44. The dragline carriage should be in line with the trench centerline. This is called the *in-line approach (Figure 7-10)*. The dragline excavates to the front while moving backwards and dumping on either side of the excavation. To ensure drainage during construction, always start at the lower end of the trench.

Sloping an Embankment

7-45. An effective use of a dragline is to dress the face of an embankment by working from the bottom to the top. Position the machine on the top of the embankment with the tracks parallel to the working face. This is called the *parallel approach (Figure 7-11)*. This positioning enables the machine to move the full length of the job without excessive turning.

Digging Underwater (Dredging) or in Wet Materials

7-46. A dragline is ideal for removing materials from areas such as water-filled trenches, canals, gravel pits, or ditches. Digging underwater or in wet materials increases the material's weight and frequently prevents the hoisting of heaped bucket loads. Plan operations so that the material being handled is as dry as possible. Always provide good project drainage. Drainage projects involving ditch excavation through swamps or soft terrain are common. Under these conditions, cast the excavated material onto a levee or spoil bank, which eliminates the problem of constructing roads for hauling-type equipment. It may be necessary to construct rudimentary service roads to support the construction effort and to get fuel to the machine.

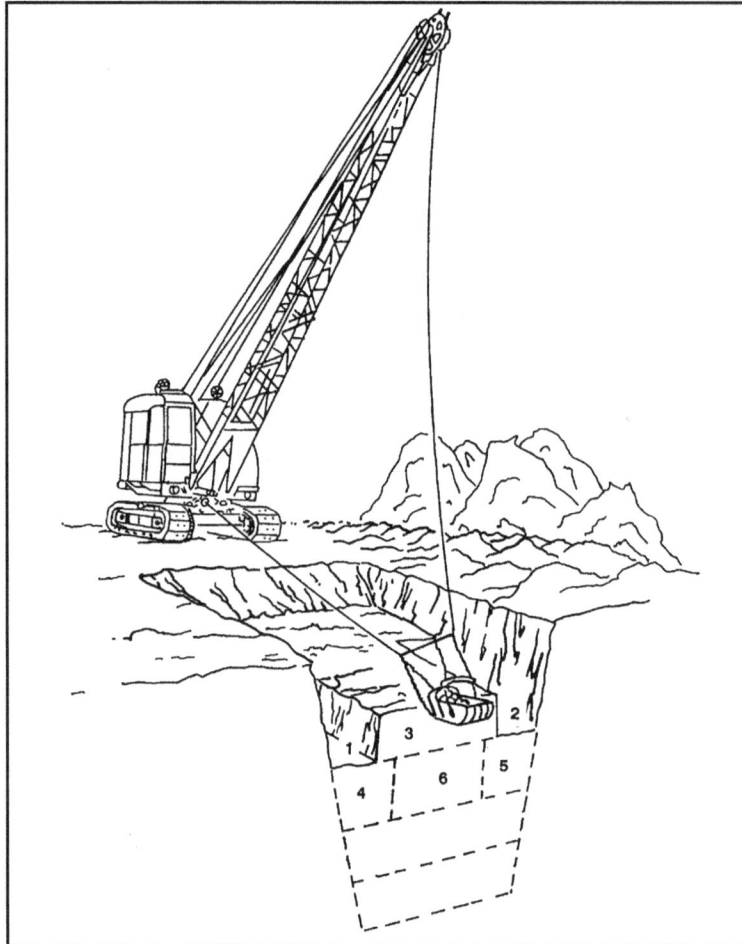

Figure 7-10. In-Line Approach With a Dragline

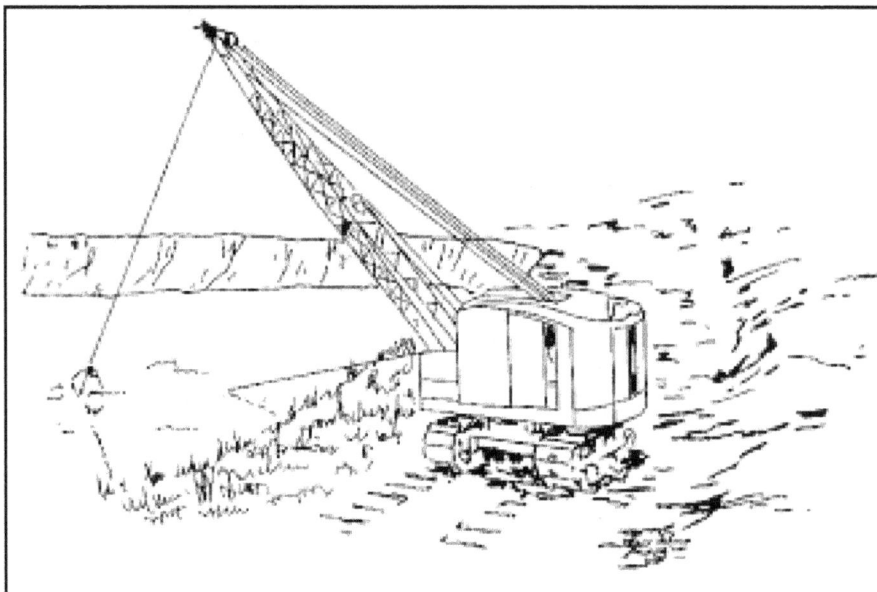

Figure 7-11. Parallel Approach With a Dragline

Loading Haul Units

7-47. Where job conditions require loading excavated material into hauling units, plan the excavation so that the loaded trucks can travel on dry ground and over minimum grades when exiting the loading area. Spot trucks for minimum boom swing. If possible, spot the truck bed under the boom tip with the truck's long axis parallel to the long axis of the boom. However, it is more common to have to spot the truck at a right angle to the boom. Spotting the haul units in the excavation below the dragline will reduce hoist time and increase production. The dragline, not being a rigid attachment, will not dump its material as accurately as other excavators. Therefore, the operator will need more time to spot the drag bucket before dumping.

PRODUCTION ESTIMATES

7-48. *Table 7-2* gives the hourly production rates for cranes with a dragline attachment. These rates are based on the optimum cutting depth, a 90 ° swing angle, the soil type, and the maximum efficiency. *Table 7-3* gives correction factors for different depths of cut and swing angles. Refer to *Table 1-1, page 1-4,* for soil conversion factors. Determine overall efficiency from past experiences.

Table 7-2. Dragline Hourly Output in BCY

Material	Dragline Bucket Size (Cubic Yards)								
	3/8	1/2	3/4	1	1 1/4	1 1/2	1 3/4	2	2 1/2
Clay or loam (light and moist)	5.0 70.0	5.5 95.0	6.0 130.0	6.6 160.0	7.0 195.0	7.4 220.0	7.7 245.0	8.0 265.0	8.5 305.0
Sand or gravel	5.0 65.0	5.5 90.0	6.0 125.0	6.6 155.0	7.0 185.0	7.4 210.0	7.7 235.0	8.0 255.0	8.5 295.0
Good common earth (soil)	6.0 55.0	6.7 75.0	7.4 105.0	8.0 135.0	8.5 165.0	9.0 190.0	9.5 210.0	9.9 230.0	10.5 265.0
Clay (hard and tough)	7.3 35.0	8.0 55.0	8.7 90.0	9.3 110.0	10.0 135.0	10.7 160.0	11.3 180.0	11.8 195.0	12.3 230.0
Clay (wet and sticky)	7.3 20.0	8.0 30.0	8.7 55.0	9.3 75.0	10.0 95.0	10.7 110.0	11.3 130.0	11.8 145.0	12.3 175.0

NOTE: The top figures give the optimum depth of cut (in feet). The bottom figures give the estimated BCY per hour.

Table 7-3. Depth-of-Cut and Swing-Angle Correction Factors for Dragline Output

Depth of Cut (in Percent of Optimum)	Swing Angle (Degrees)							
	30	45	60	75	90	120	150	180
20	1.06	0.99	0.94	0.90	0.870	0.81	0.75	0.70
40	1.17	1.08	1.02	0.97	0.930	0.85	0.78	0.72
60	1.24	1.13	1.06	1.01	0.970	0.88	0.80	0.74
80	1.29	1.17	1.09	1.04	0.990	0.90	0.82	0.76
100	1.32	1.19	1.11	1.05	1.000	0.91	0.83	0.77
120	1.29	1.17	1.09	1.03	0.985	0.90	0.82	0.76
140	1.25	1.14	1.06	1.00	0.960	0.88	0.81	0.75
160	1.20	1.10	1.02	0.97	0.930	0.85	0.79	0.73
180	1.15	1.05	0.98	0.94	0.900	0.82	0.76	0.71
200	1.10	1.00	0.94	0.90	0.870	0.79	0.73	0.69

EXAMPLE

Determine the hourly output for a 3/4-cubic-yard crawler dragline.

Bucket size = 3/4 cubic yard
Material = good common earth
Angle of swing = 45°
Depth of cut = 9 feet

Step 1. Determine the ideal production from *Table 7-2.*

Ideal production = 105 BCY per hour at an optimum depth of cut of 7.4 feet

Step 2. Determine the ratio of the actual depth of cut to the optimum depth of cut, expressed as a percent.

Cut ratio (percent)

$$= \frac{\text{actual depth of cut}}{\text{optimum depth of cut}} \times 100$$

$$= \frac{9 \text{ feet}}{7.4 \text{ feet}} \times 100$$

$$= 122 \text{ percent of optimum}$$

Step 3. Determine the depth-of-cut/swing-angle correction factor from *Table 7-3.* In some cases it may be necessary to interpolate between *Table 7-3* values.

Correction factor = 1.17

Step 4. Determine an overall efficiency factor based on the job conditions. Draglines seldom work at better than a 45-minute working hour.

$$\text{Efficiency factor} = \frac{45 \text{ minutes}}{60 \text{ minutes}} = 0.75$$

Step 5. Determine the production rate. Multiply the ideal production by the depth-of-cut/swing-angle correction factor and the efficiency factor.

Production rate = 105 BCY per hour $\times$ 1.17 $\times$ 0.75 = 92.1 BCY per hour

Step 6. Determine the soil conversion factor, if needed.

Soil conversion factor = not applicable

OPERATION TIPS

- Position the machine to eliminate unnecessary casting and hoisting, although the dragline bucket can easily be cast beyond the length of the boom.
- Use heavy timber mats for work on soft ground. Keep the mats as level and clean as possible.

WARNING
Do not guide the dragline bucket by swinging the superstructure while digging. This puts side stress on the boom, which can cause the boom to collapse. Raise the bucket clear of the ground before swinging the boom.

SAFETY

OPERATOR RESPONSIBILITIES

7-49. Thoroughly train operators in crane safety before allowing them to operate a crane. Operators are responsible for knowing the limitations and capabilities of cranes and how to read a load chart properly. Operators must not operate an unsafe crane. They must be able to identify and promptly report any equipment malfunction or defect. Operators have the authority to stop and refuse to handle loads until safety has been assured.

HAND SIGNALS

7-50. Use a signal person whenever the point of operation is not in full and direct view of the equipment operator. When using hand signals, designate only one person to give the signals to the operator. The signal person must be totally dependable and fully qualified. The signal person must use a uniform system of signals and must be clearly visible to the operator at all times. *Figure 7-12* shows the hand-signal system recommended for directing crane-shovel operations. Check the operator's manual for other equipment signals. Ensure that operators and signalmen have a full understanding of the meaning of all signals. See *Chapter 13* for additional crane-safety precautions.

ACCIDENT PREVENTION

7-51. Common hazards associated with operating hook blocks, clamshells, pile drivers, and draglines are—

- The boom contacting high-voltage electric wires. This is the most hazardous aspect of crane operation.
- The cables breaking.
- The clutch or brake slipping, allowing the boom radius to increase.
- Obstruction of the free passage of the boom or the load.
- Operation on uneven ground.
- Not knowing the actual weight of the load being hoisted.
- Bent or dented chord members on the boom.

7-52. Common operator errors associated with crane operations are—

- Dropping or slipping the load.
- Not using outriggers.
- Not using mousing or safety-type hooks.
- Not being familiar with the equipment.
- Not referring to load charts when using different boom lengths.
- Using the crane hoist cable for towing. The boom is designed for handling vertical loads only.

7-53. Personnel should never attempt to climb on or off of a crane when it is operating. No personnel except the operators and, on occasion, examiners, supervisors, trainees, or repairers, should be on a crane while it is in operation.

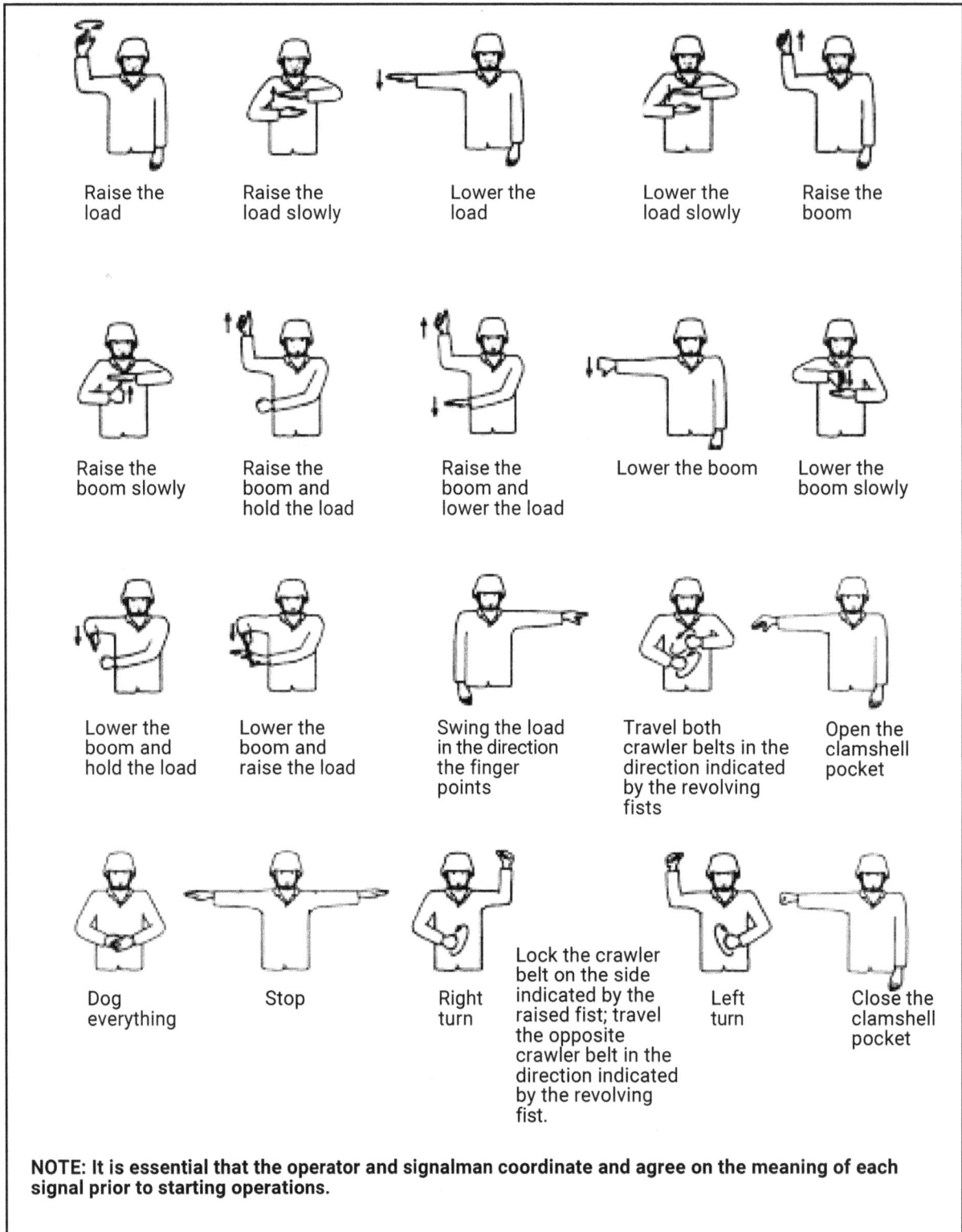

Raise the load

Raise the load slowly

Lower the load

Lower the load slowly

Raise the boom

Raise the boom slowly

Raise the boom and hold the load

Raise the boom and lower the load

Lower the boom

Lower the boom slowly

Lower the boom and hold the load

Lower the boom and raise the load

Swing the load in the direction the finger points

Travel both crawler belts in the direction indicated by the revolving fists

Open the clamshell pocket

Dog everything

Stop

Right turn

Lock the crawler belt on the side indicated by the raised fist; travel the opposite crawler belt in the direction indicated by the revolving fist.

Left turn

Close the clamshell pocket

NOTE: It is essential that the operator and signalman coordinate and agree on the meaning of each signal prior to starting operations.

Figure 7-12. Hand Signals for Crane-Shovel Operations

7-54. When hoisting a load from below water, the crane takes on the added load imposed by the displaced water as the load is hoisted out of the water. Never hoist unknown weights from the water. Consider the water contained in the load or in a waterlogged structure as part of the load's weight. Never handle waterlogged loads or loads from water or mud without first determining whether the weight of the load and the water are within the crane's hoisting capacity.

7-55. When handling a heavy load, raise it a few inches to determine whether there is undue stress on any part of the sling and to ensure that the load is balanced. If anything is wrong, lower the load at once and do not attempt to move it until the necessary adjustment or repair has been made.

7-56. Before hoisting a near-capacity load, make sure the hoisting line is vertical. Move the crane instead of lowering the boom, since swinging a capacity load increases the chance of tipping.

7-57. When lowering a boom under load, use extreme caution. Check the load chart with attention to radius changes and observe the radius indicator. These charts are posted in the operator's cab. Never lower the hoisting line and the boom simultaneously.

7-58. When lowering loads, use a low speed not to exceed the hoisting speed of the equipment for the same load. The ordinary hoisting speed of a 30-ton, motor-operated crane is about 18 feet per minute with a rated load. Stopping the load at such speeds in a short distance may double the stress on the slings and crane.

7-59. Be careful to guard workers, buildings, or scaffolds against injury from swinging loads. Do not swing loads over workers. If it is necessary to move loads over occupied areas, give adequate warning (by bell or siren) so workers can move into safe locations.

7-60. Do not attempt dual lifts unless absolutely necessary and only with competent supervision throughout the operation. Dual lifts are extremely dangerous. Shifting of the load can cause overloading and failure of one crane. This throws the entire load onto the second crane causing it to fail. Before making a dual lift, carefully determine the position for the cranes and the location of the slings to balance the load properly.

7-61. After repair or alteration of a crane or derrick involving its hoisting capacity or stability, have a competent person determine its safe working load. Have this person issue a written statement specifying the safe working load.

7-62. Test the brakes at the beginning of each new shift, after a rainstorm, or at any other time when brake linings may have become wet. When hoisting a capacity load, check the brakes by stopping the hoist a few inches above the ground and holding it with the brake.

7-63. Equip all cranes with appropriate fire extinguishers. Keep the extinguishers maintained and ready for use.

7-64. Never attempt to pull pipes or other objects out of the ground.

PILE-DRIVER SAFETY

7-65. When positioning a crane to drive piles, prepare a level work platform and use outriggers to maintain crane stability. When in operation, use safety lashings for all hose connections to pile drivers, pile ejectors, or jet pipes. Use tag lines to control piles and hammers. When hoisting steel piling, use a closed shackle or other positive means of attachment. Only the pile-driving crew members should be permitted in the work areas when driving piles. Pile-driver operators should be aware of the following safety precautions:

- **Repairs.** Never repair any diesel or air equipment while it is in operation or under pressure.
- **Air hoses.** Make frequent inspections of air hoses to locate defects and promptly replace any defective air hoses.
- **Power lines.** Make sure that machines or equipment do not come too close to electric power lines. The machine does not actually have to contact a power line for the machine to be energized.
- **Hammer and driving heads.** When a pile driver is not in use, use a cleat or timber to hold the hammer in place at the bottom of the leads. Secure the driving heads when using the rig to shift cribbing or other material. Never place your head or other parts of your body under suspended hammers that are not dogged or blocked in the leads.
- **Drums, brakes, and leads.** Keep hoisting drums and brakes in the best condition possible and shelter them from the weather. Keep leads well greased to provide smooth hammer travel.

Chapter 8
Hydraulic Excavators

Hydraulic excavators are designed to excavate below the ground surface on which the machine rests. These machines have good mobility and are excellent for general-purpose work, such as excavating trenches and pits. Because of the hydraulic action of their stick and bucket cylinders, they exert positive forces crowding the bucket into the material to be excavated. The major components of the hydraulic hoe are the boom, the stick (arm), and the bucket.

DESCRIPTION

8-1. Fast-acting, variable-flow hydraulic systems and easy-to-operate controls give hydraulic excavators high implement speed and breakout force to excavate a variety of materials. The hydraulic hoe is ideal for excavating below the ground's surface on which the machine rests. A large variety of booms, sticks, buckets, and attachments give these excavators the versatility to excavate trenches, load trucks, clean ditches, break up concrete, and install pipes. The small emplacement excavator (SEE) with its hoe attachment *(Figure 8-1)* can work in tight places and has good mobility.

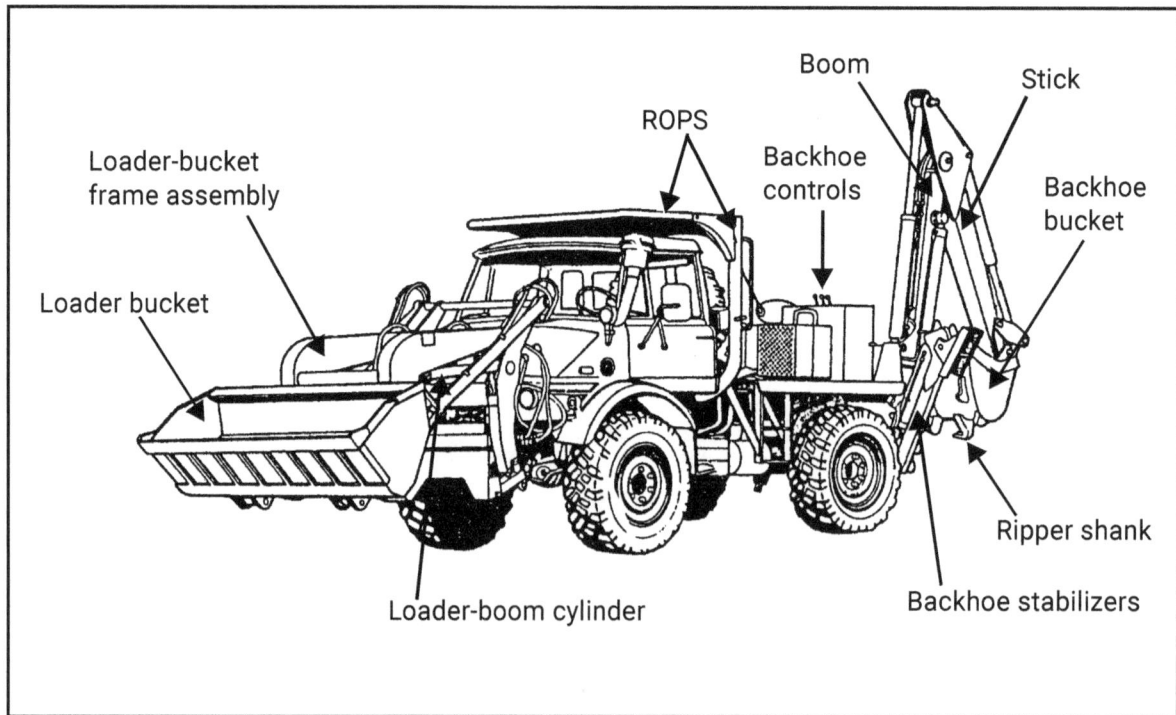

Figure 8-1. Small Emplacement Excavator

EXCAVATION TECHNIQUES

8-2. The hoe is normally associated with two types of excavations—trenching (linear-type) and basement (area-type). The operator should judge the length and depth of cut to produce a full bucket with every pass*(Figure 8-2)*.

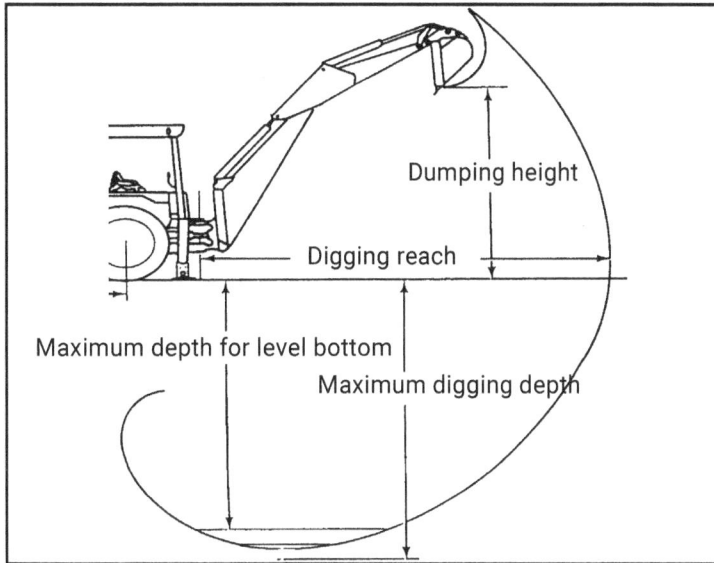

Figure 8-2. Hoe-Bucket Operating Dimensions

TRENCHES

8-3. *Figure 8-3* shows parallel and perpendicular trenching using a SEE hoe attachment.

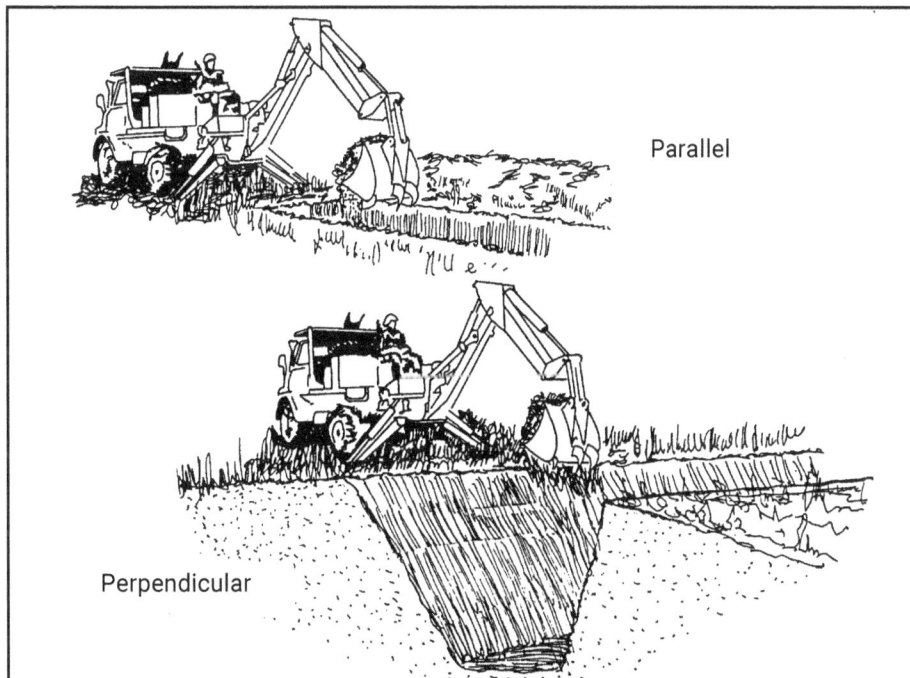

Figure 8-3. A SEE Digging Trenches With a Hoe Attachment

Parallel

8-4. With the parallel method, center the hoe on the trench, while keeping the tractor in line with the trench center line. As the digging progresses, move the machine away from the excavation and load the material into haul units or stockpile it along the side of the trench for later use as backfill.

Perpendicular

8-5. When using the perpendicular method, dig the trenches in two or more cuts or lifts. To excavate the top 35 to 45 percent of the trench depth, make the first cut with the boom carried high. To finish the cut and remove the remainder of the material, move forward about one-half the length of the machine with the boom carried low. Although this method involves more and shorter moves, it has better bucket digging angles and shorter hoisting distance on the top lifts.

BASEMENTS

8-6. Many variations of the two basement-excavation sequences shown in *Figure 8-4* are possible. The procedures vary with the design and shape of the excavation, the restrictions of surrounding properties, and the requirements for disposing of the spoil. The first cut is a trench with vertical outside walls. To minimize handwork or cleanup, dig all outside wall faces vertically. Plan the starting point and the digging sequence so that the machine conveniently works itself out of the excavation. Dig trenches for service pipes last; dig them from the basement outward. Straddle the machine over the outer edge and dig over the end and side of the tractor. Move the machine as the arrows in the figure indicate.

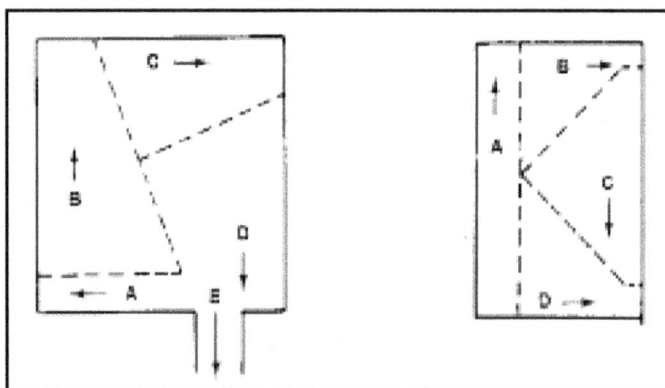

Figure 8-4. Two Methods of Excavating Basements

OPERATION TECHNIQUES

UNDERGROUND UTILITIES

8-7. Survey the area for underground hazards as well as for surface obstacles before digging. This applies particularly to populated areas with multiple underground utilities.

CONFINED QUARTERS

8-8. Working in confined quarters is not efficient from a production standpoint. If expecting considerable close-quarter work, plan to use small machines that can operate efficiently with a minimum work radius.

DRAINAGE DITCHES

8-9. If the job is to continue during wet seasons or in wet areas, give prime consideration to drainage. Begin ditch excavations at the lower end and work toward the upgrade.

HARD MATERIALS

8-10. A hoe will dig into fairly hard materials. However, blasting or ripping may be more efficient than breaking through hardpan and rock strata with the bucket. Once the trench is open, break the ledge rock by pulling the bucket up under the layers. Remove the top layers first, lifting only one or two layers at a time.

SMALL EMPLACEMENT EXCAVATOR WITH A LOADER BUCKET

8-11. The SEE is a lightweight, all-wheel-drive, diesel-engine, high-mobility machine. It is equipped with a hoe, a loader bucket, and other hydraulic attachments, which normally include a hammer drill, a chainsaw, and a pavement breaker. Check the operator's manual for using the SEE's hydraulic-mounted attachments. The SEE weighs less than 16,000 pounds, is air-transportable, can travel more than 50 mph on improved roads, and has excellent off-road mobility.

EXCAVATING

8-12. Under average conditions, small hoes—bucket size less than 1 cubic yard—can complete an excavation cycle in 14 seconds. An excavation cycle consists of loading the bucket, swinging the loaded bucket, dumping the bucket, and swinging the empty bucket. Average conditions would be a depth of cut between 40 and 60 percent of the machines rated maximum digging depth and a swing angle of between 30 ° to 60°. The average cycle time for bucket sizes from 1 cubic yard to less than 2.5 cubic yards is 15 seconds. Greater digging depths or swing angles increase the cycle time.

8-13. Make sure the hoe is level before operating. Lower the front loader bucket to the ground (flat) so that the machine's front wheels are not in contact with the surface. Move the gearshift and the range-shift levers to their neutral positions, and lower the outriggers. Use the outriggers to level the machine and to raise the rear wheels slightly above the ground. Always operate with the least amount of bucket-arm swing.

8-14. For evaluating heaped capacity, hoe buckets are rated with an assumed material repose angle of 1:1. Therefore, actual bucket capacity depends on the type of material being excavated as all materials have their own natural repose angle. *Table 8-1* provides bucket fill factors for hoe buckets based on material type.

Table 8-1. Bucket Fill Factors for Hoe Buckets

Material	Fill Factor (Percent)
Moist loam or sandy clay	100 to 110
Sand and gravel	95 to 110
Rock (poorly blasted)	40 to 50
Rock (well blasted)	60 to 75
Hard, tough clay	80 to 90

8-15. **Bucket Cylinder.** When using the bucket cylinder to excavate, follow these steps *(Figure 8-5)*:

Step 1. Put pressure on the boom to force the bucket teeth or cutting edge into the ground.

Step 2. Roll the bucket toward the machine until it is full.

Step 3. Raise the bucket, in a smooth operation, high enough above the trench to clear the spoil pile or the hauling unit, and dump the excavated material.

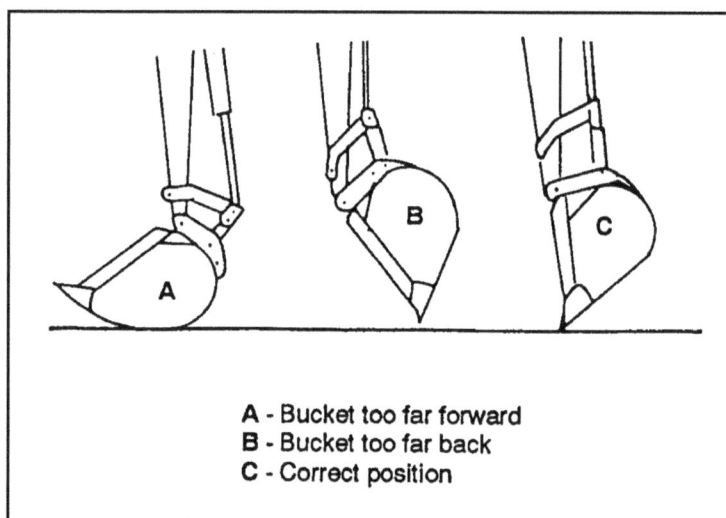

A - Bucket too far forward
B - Bucket too far back
C - Correct position

Figure 8-5. Bucket-Cylinder Operation

8-16. **Stick Cylinder.** The maximum digging force is developed by operating the stick cylinder perpendicular to the stick. As a rule, the optimum depth of cut for a hoe is 30 to 60 percent of the machine's maximum digging depth *(Figure 8-2, page 8-2)*. When using the stick cylinder to excavate, follow these steps:

Step 1. Lower the bucket into the digging position *(Figure 8-6 [A])*.

Step 2. Roll the bucket until the bucket teeth or the cutting edge is flat on the ground *(Figure 8-6 [B])*.

Step 3. Use the stick cylinder to move the bucket toward the machine until it is half full *(Figure 8-6 [C]).*

Step 4. Raise the stick and roll the bucket until it is full*(Figure 8-6 [D]).*

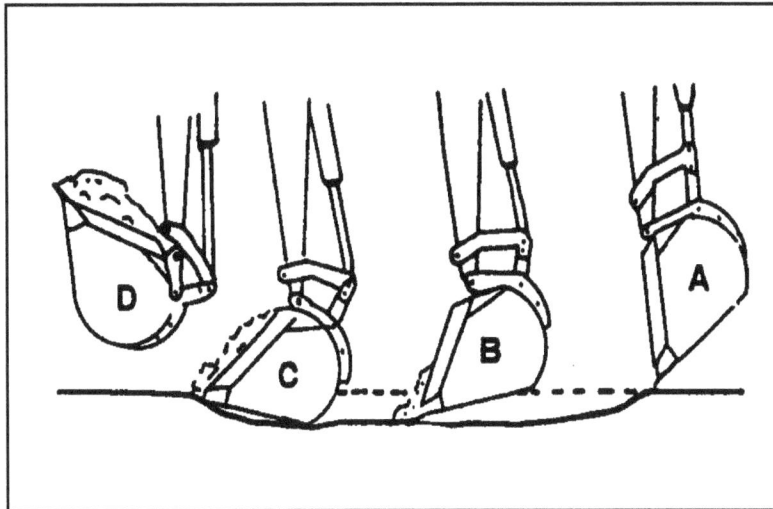

Figure 8-6. Hoe Digging Technique

Loader Bucket

8-17. When digging with a loader bucket—

- Use the bucket cylinders to help break the ground loose instead of depending on the forward movement of the machine, as in the loader crowding technique.
- Do not raise the bucket higher than necessary to dump the material.
- Use as flat a ramp as possible when starting an excavation. Plan the job so that most of the hauling from the excavation can be done when driving the unit forward. A steeper ramp can be used when driving forward than when driving in reverse.
- Keep the working area level.

LOADING

8-18. To excavate and load, a hoe bucket must raise through the digging motion and above the haul unit. If possible, spot the truck on the pit floor. The bucket will then be above the haul unit when the digging is complete. At that point it is not necessary to raise the bucket further before swinging and dumping. This arrangement will save about 12 percent of the total excavation-loading cycle time. When loading dump trucks with a hoe —

- Plan and lay out the area of operation.
- Spot the truck so that the hoe does not have to turn (revolve) more than 90° (V-positioning, discussed in *Chapter 10,* is often appropriate).

- Rotate the bucket over the rear of the dump bed, rather than over the cab of the truck.
- Keep the working area smooth.
- Raise the bucket while moving toward the truck.
- Lower the bucket while moving away from the truck.
- Shake the bucket only when necessary to loosen dirt stuck in the bottom of the bucket.

LEVELING AND GRADING

8-19. Use the loader bucket for leveling and grading, as follows:

- Fill all holes and hollows and loosen up any high spots before attempting to finish the grade *(Figure 8-7)*.
- Spread the dirt evenly by holding the bucket close to the grade (tipped slightly forward) and letting the dirt spill.
- Level and pack the dirt with the loader bucket in a lowered position. To finish, operate the machine in reverse with the bucket dragging (back blading) on the ground.

Figure 8-7. Leveling the Ground With a Loader Bucket

TRENCH BACKFILLING

8-20. Use the loader bucket for trench backfilling as follows:

- Position the machine at approximately a 45° angle to the length of the trench and its spoil pile.
- With the bucket raised about 2 inches above the natural ground, use it like a dozer blade to push the material into the trench. Keep the bucket level while pushing the material; do not crowd/curl.

- After the material falls into the trench, reverse the machine and move along the pile to repeat pushing.
- After the last pass, dump the material remaining in the bucket into the trench.

If the material in the spoil pile along the trench is higher than 2 feet or is wet, attack the pile in two passes. Take off the upper half with the first pass and the remainder with a second cleanup pass.

TRACK-MOUNTED EXCAVATOR

8-21. Track-mounted excavators *(Figure 8-8)* are diesel-engine machines that have a maximum digging depth of approximately 20 feet and an approximate dumping height of 22 feet. These excavators can travel around a job site at a maximum speed of about 3 mph in high range. They must be transported for long-distance travel between projects. They are used for excavating pipeline trenches, drainage ditches, building footings, and hasty fortifications and for loading trucks.

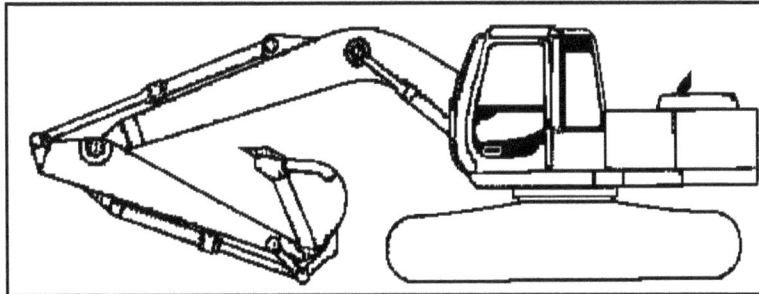

Figure 8-8. Track-Mounted Excavator

EXCAVATING

8-22. These excavators can be equipped with buckets ranging in size from 1 to 2.5 cubic yards. The excavation cycle for this machine is about 15 seconds based on average conditions, a depth of cut between 8 and 12 feet, and a swing angle of 30 to 60°.

LIFTING

8-23. On utility jobs the excavator many need to lift and swing heavy sections of pipe into a trench. Sometimes these machines are used to hoist and unload materials from trucks. The weight an excavator can lift depends on the distance the load is from the center of gravity of the machine. Always refer to the current specification sheets before attempting a lift. Position the machine as close to the load as possible. The other critical element to consider is swing and position. The lifting capability is 65 to 70 percent greater over the front of the machine than over the side. These machines are designed to handle 15,000 pounds (at a swing radius of 15 feet) over the side.

PRODUCTION ESTIMATES

8-24. Factors that affect hoe production are the—

- Width of the excavation.
- Depth of the cut.

- Material type.
- Working radius for digging and dumping.
- Required bucket dumping height.

EXAMPLE

Use a hoe equipped with a 0.25-cubic-yard bucket to excavate hard clay. The depth of cut will average about 50 percent of the machine's maximum digging depth and the swing angle should be less than 60°. What is the expected production, in BCY per hour, assuming 50 working minutes per hour?

Step 1. Determine the bucket fill factor based on the material type *(Table 8-1, page 8-5).*

Fill factor for hard clay = 80 to 90 percent

Lacking any other information, use an average of 85 percent.

Step 2. Use a hoe cycle time based on past performance data if available or use the average cycle time given in *paragraph 8-12.*

Average cycle time = 14 seconds

Step 3. Determine the ideal production rate (LCY per hour).

Ideal production rate (LCY per hour) =

$$\frac{3,600 \text{ seconds per hour}}{\text{backhoe cycle time (seconds)}} \times \text{bucket size (cubic yards)} \times \text{fill factor (from step 1)}$$

Ideal production rate $= \frac{3,600}{14} \times 0.25 \times 0.85 = 55$ **LCY per hour**

Step 4. Determine the production rate (LCY per hour) by adjusting for efficiency.

Production rate = ideal production rate (LCY per hour) × efficiency factor

Production rate (LCY per hour) $= 55 \times \frac{50 \text{ minutes}}{60 \text{ minutes}} = 45$ **LCY per hour**

Step 5. Convert the production rate from LCY per hour to BCY per hour. Determine the soil-volume correction factor from *Table 1-1, page 1-4* (LCY to BCY for hard clay).

Soil conversion factor for clay (loose to bank) = 0.7

Production rate = 45 LCY per hour × 0.7 = 32 **BCY per hour**

Chapter 9

Air Compressors and Pneumatic Tools

When air is compressed, it receives energy from the compression. This energy is transmitted through a pipe or a hose to the operating equipment, where a portion of the energy is converted into mechanical work. A compressed-air system consists of one or more compressors together with a distribution system to carry the air to the points of use. Engineer units use compressed air to inflate rubber equipment, to spray materials, to operate pneumatic tools, to clean equipment, and to perform certain jobs in maintenance shops. A pneumatic tool uses the energy of compressed air as the power for its operation.

AIR COMPRESSORS

9-1. Portable air compressors are commonly used on construction sites where it is necessary to meet frequently changing job demands. *Figure 9-1* and *Figure 9-2, page 9-2,* show two types of air compressors. The capacity of an air compressor is determined by the amount of free air that it can compress to a specified pressure in one minute, under standard conditions (absolute pressure of 14.7 pounds per square inch [psi] at 60°F). This amount of free air is usually expressed in cubic feet per minute (cfm). The number of pneumatic tools that can be operated from one air compressor depends on the air requirements of the specific tools.

Figure 9-1. Trailer-Mounted, 250-cfm Air Compressor

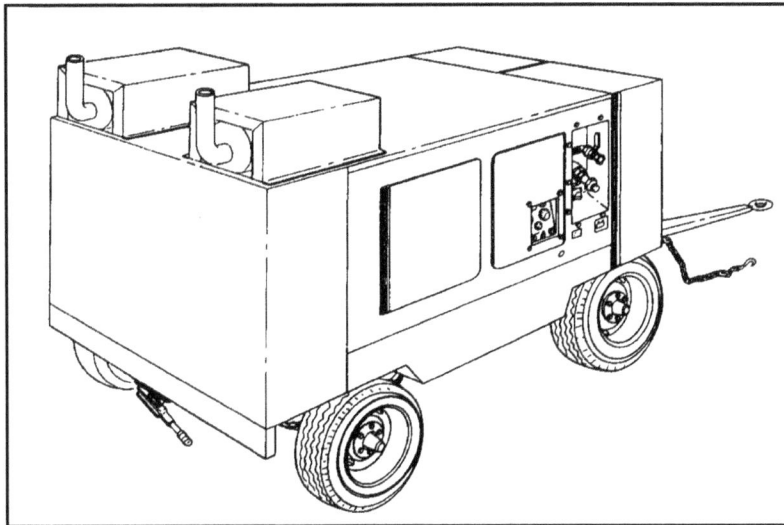

Figure 9-2. Wheel-Mounted, 750-cfm Air Compressor

EFFECTS OF ALTITUDE

9-2. When a given volume of free air is compressed, the original pressure will average 14.7-psi absolute pressure at sea level. If the same volume of free air is compressed to the same gauge output pressure at a higher altitude, the volume of the air after being compressed will be less than the volume compressed at sea level. Thus, while a compressor may compress air to the same discharge pressure at a higher altitude, the volume supplied in a given time interval will be less at the higher altitude. *Table 9-1* shows the percentage of volumetric efficiency at different altitudes based on a 100-psi gauge output pressure.

Table 9-1. Efficiency of Air Compressors at Various Altitudes (100-psi Gauge Output Pressure)

	Single-Stage Reciprocating Compressor	Two-Stage Reciprocating Compressor	Rotary Compressor
Altitude (Feet)	Percent of Efficiency	Percent of Efficiency	Percent of Efficiency
2,000	98.7	99.4	100.0
5,000	92.5	98.5	100.0
7,000	–	–	100.0
8,000	87.3	97.6	99.9
10,000	84.0	97.0	–
12,000	–	–	98.6

CAPACITY OF COMPRESSORS

9-3. On a typical job, some tools operate almost continuously, while others operate infrequently. An analysis should be made to determine the probable *actual need* before determining the compressor requirements. If ten rock drills are nominally drilling, generally no more than five or six of the drills will be consuming air at a given time. Additionally, the amount of air used will vary considerably in different applications. *Table 9-2* provides data on the air requirements of specific tools. After considering the number of working tools and their air requirements, increase the total amount of air demanded by 10 percent to compensate for leakage.

Table 9-2. Description and Operating Data for Pneumatic Tools

Tool	Air Requirements		Hourly Work Output	Air-Line Hose Diameter
	psi	cfm		
Paving breaker (jackhammer) (80-lb) Attachments: chisel point, moil point, 8" diameter tamper, sheeting driver	80-100	65	15 sq ft of 6" asphalt paving with chisel point; 12 to 50 sq ft of 6" to 8" nonreinforced concrete with moil point depending upon width of cut (12 sq ft for narrow cut, 50 sq ft for wide cut); maximum lift of 8" with tamper; driving wood or up to 2" thick steel sheeting with sheeting driver	3/4"
Paving breaker (clay digger) (25-lb) Attachments: moil point, pick, spade, drum opening tool	80-90	35	Loosen 1.2 cubic yards of tough clay with spade; open 20 to 30, 55-gallon drums with a drum-ripping tool	1/2"
Nail driver Attachments: 1/2" and 3/4" head driving sets, rivet buster	90	30	250, 80d nails (after the nails have been started with a hammer or a sledge)	1/2"
Circular saw Attachments: woodworking, 12" blade (Special abrasive disks available for cutting other materials.)	80-100	75	Cuts 4" x 4" lumber in 30 seconds	1/2"
Chain saw Attachments: 24" blade, Type I, Size 1	80-100	90	Cuts 12" diameter hardwood log in 50 seconds. (Under 25': 5/8", 25' to 100': 3/4", over 100': 1")	5/8" to 1/2"
Wood drill Attachments: 2" diameter capacity. Ship augers range from 7/16" to 2" in 1' and 3' lengths.	80-100	60	Using largest size auger, 125 holes 36" deep	3/4"
Sump pump (3" discharge, 175-GPM capacity)	80-90	100	175 GPM at 25' head 150 GPM at 150' head	3/4"
Steel drill (Number 3 Morse taper chuck, 1 1/4" capacity)	90-100	27	30 each 1" diameter holes in 1" thick steel plate (lead holes of 1/4" diameter drilled previously)	1/2"
Handheld rock drill (dry-type, 55-lb) Attachments: 2', 4', 6', and 8' hollow-steel drill rods and 1 5/8, 1 3/4", 1 7/8", and 2" drill bits	80-100	95	1 3/4" Hole Soft rock: 15-20' Medium rock: 10-15' Hard rock: 5-10'	3/4"

NUMBER OF COMPRESSORS

9-4. The number of compressors required will depend on the sizes available. Normally, if 1,400 cfm of free air were required for a specific job, two 750-cfm units would be sufficient. Air compressors are sturdy machines, but like all mechanical equipment they require maintenance. Therefore, in some cases a standby unit will be required.

LOCATION OF COMPRESSORS

9-5. If possible, centrally locate all compressors on the job. This arrangement has the advantage of unified operation and better supervision. It is possible that a central location is not advisable due to a lack of piping, too large a friction loss, or obstructions on the job site. In this case, it would be necessary to locate compressors at appropriate points. All air compressors must be leveled and should be placed as close to the air-operated devices as conditions will permit.

OPERATION OF COMPRESSORS

9-6. Air compressors should always be located upwind from the work to keep foreign material out of the air intake. When operating under extremely dusty conditions, take precautions to protect the units from as much dust as possible. Other factors to consider are as follows:

- Open all drain cocks to drain condensation after each 8 hours of operation, thus eliminating the possibility of rusting or freezing.
- Close the side panels of the compressor housing when it is being operated in cold weather.
- Block the wheels and engage the hand brake of the trailer mount before operation.
- Ensure that the receiver tank is drained of air when operations are complete.

COMPRESSED-AIR USES

9-7. Compressed air is used extensively on construction projects. In many instances, compressed air is the most convenient method of operating equipment and tools.

ASPHALT PLANTS

9-8. Air compressors are frequently used in asphalt plants for fuel atomization of the dryer burner. Compressors are also used to clean up the plant, to power various tools at the paving site, and to dedrum.

CONCRETE OPERATIONS

9-9. At the batch plant, vibrators may be used on the aggregate hopper to prevent bridging. Air-driven pin drivers and cleaning devices for cleaning sawed joints are used at the paving site.

PNEUMATIC TOOLS

9-10. The military uses pneumatic paving breakers, nail drivers, saws, drills, pumps, and a variety of other pneumatic tools. Pneumatic tools can be powered by either a reciprocating-percussion or a rotary-vane air motor.

- **Reciprocating-percussion air motor.** The reciprocating-percussion air motor is used when a hammering action is desired. It employs a free-floating piston moving in a cylinder. When the throttle is opened, a set of valves introduce air alternately to the ends of the cylinder, driving the piston back and forth. The force of the piston is transmitted to the tool, which does the work.
- **Rotary-vane air motor.** The rotary-vane air motor is employed when a rotary motion is desired. The motor employs a cylinder having an eccentrically mounted slotted rotor, with each slot containing a spring-loaded vane. When the throttle is opened, compressed air enters a small compartment. Pressure on the vanes causes the rotor to turn in the direction of a larger compartment. A gear train transmits the rotation to the attachment, which does the work.

9-11. Two important factors that affect the condition of a pneumatic tool are lubrication and air pressure.

- **Lubrication.** To check for proper lubrication of a pneumatic tool, pass a piece of paper in front of the tool exhaust port. If a thin film of oil accumulates on the paper, the tool is being properly lubricated. If drops of oil appear on the paper or if oil foams around the exhaust port, the tool is over-lubricated. If no oil appears, the lubrication device should be checked immediately.

- **Air pressure.** Each tool requires a specified volume of air at a specific pressure. If the volume of air or pressure is allowed to drop excessively, considerable damage will result. Check for air leaks in the hose and around the air connections. Listen to and observe the tool when it is operating. If a tool appears to be operating sluggishly or appears to be surging (erratic operation), it has either too much or too little pressure. The tools should never be operated with less than 70- or more than 100-psi pressure at the tool. If the air-pressure gauge on the air compressor continually remains below 70 psi, the unit is overloaded (too little pressure at the tool).

9-12. Most pneumatic tools are heavy and create a considerable amount of vibration. A difficulty sometimes encountered with their use is operator fatigue. This is a particular problem with inexperienced operators. Careful attention should be given to the selection of operators to ensure that they are in good physical condition and strong enough to operate the equipment.

AIR MANIFOLDS

9-13. Many construction jobs require more compressed air per minute than any one compressor will produce. An air manifold is a large-diameter pipe used to transport compressed air from one or more air compressors without a detrimental friction-line loss.

CONSTRUCTION

9-14. Manifolds can be constructed of any durable pipe. Compressors are connected to the manifold with flexible hoses. A one-way check valve must be installed between the compressor and the manifold. This valve keeps the manifolds back pressure from possibly forcing air back into a compressor's receiver tank. The compressors that are grouped to supply an air manifold may be of different capacities, but the final discharge pressure of each should be coordinated at 100 psi. Compressors of different types should not be used on the same air manifold. The difference in the pressure control systems of a rotary and a reciprocating compressor could cause one compressor to become overloaded while the other compressor idles. The Army commonly constructs air manifolds of 6-inch-diameter *invasion* piping. Air may be used at any point along the manifold by installing outlet valves for connecting air lines and pneumatic tools.

COMPRESSED-AIR DISTRIBUTION SYSTEM

9-15. The purpose of installing a compressed-air distribution system is to provide a sufficient volume of air to the work site at pressures adequate for efficient tool operation. Any drop in pressure between the compressor and the point of use is an irretrievable loss. Therefore, the distribution system is one of the most important elements of the total system. Observe the following general rules in planning a compressed-air distribution system:

- Pipe sizes should be large enough so that the pressure drop between the compressor and point of use does not exceed 10 percent of the initial pressure.
- Each header or main line should be provided with outlets as close as possible to the point of use. This permits shorter hose lengths and avoids large pressure drops through the hose.
- Condensate drains should be located at appropriate places along the headers or main lines.

FRICTION LOSSES

9-16. The hose or pipe connecting the tool to the air compressor resists the flow of air. Consequently, the pressure at the tool end of the line is less than at the compressor end. The air-line friction increases as the diameter of the hose or pipe decreases or as the length of the hose or pipe increases. Through practice, it has been determined that a 200-foot-long, 3/4-inch-diameter hose is the maximum length and diameter to which a handheld tool can be connected and operated efficiently. Standard tables *(Tables 9-3* and *9-4)* are available for calculating the friction loss in a pipe or hose.

AIR-LINE HOSE

9-17. Air-line hose is a rubber-covered, pressure-type hose used for transmitting compressed air. Hose with a 3/4-inch inside diameter is used with hand-operated tools and hose with a 2-inch inside diameter is used with a crawler-mounted drill. Hose is usually furnished in 50-foot lengths and equipped with quick-acting fittings (for attaching a tool, a compressor, or another hose). Leader hose is made of oil-resistant neoprene rubber and has end attachments. It is used between the air-line oiler and an air tool. Sections of leader hose are usually 12 or 25 feet long and 1/2 or 3/4 inch in diameter.

AIR-LINE OILER

9-18. The air-line oiler is a reservoir which is placed in the air line directly in front of the air tool for the purpose of lubricating the tool. As the air passes through the oiler, it picks up the oil which is carried into the tool. An adjustable needle controls the amount of oil entering the air stream. There are both directional and nondirectional oilers. The arrow should be pointed in the direction of the airflow when it is connected to the air line.

PNEUMATIC TOOLS

9-19. Pneumatic tools are simpler in design than similar gasoline or electric-powered tools and require less maintenance. A pneumatic tool with nonsparking attachments can be operated around petroleum products or explosive materials without presenting a fire hazard.

Table 9-3. Loss of Air Pressure Due to Friction in a Pipe [1]
(in psi per 1,000 Feet of Pipe With 100-Pound-Gauge Initial Pressure)

Cubic Feet of Free Air Per Minute	Nominal Pipe Diameter (Inches)								
	1	1 1/4	1 1/2	2	2 1/2	3	3 1/2	4	4 1/2
100	27.9	6.47	2.86	0.77	0.30	–	–	–	–
125	48.6	10.20	4.49	1.19	0.46	–	–	–	–
150	62.8	14.60	6.43	1.72	0.66	0.21	–	–	–
175	–	19.80	8.72	2.36	0.91	0.28	–	–	–
200	–	25.90	11.40	3.06	1.19	0.37	0.17	–	
250	–	40.40	17.90	4.78	1.85	0.58	0.27	–	
300	–	58.20	25.80	6.85	2.67	0.84	0.39	0.20	–
350	–	–	35.10	9.36	3.64	1.14	0.53	0.27	–
400	–	–	45.80	12.10	4.75	1.50	0.69	0.35	0.19
450	–	–	58.00	15.40	5.98	1.89	0.88	0.46	0.25
500	–	–	71.60	19.20	7.42	2.34	1.09	0.55	0.30
600	–	–	–	27.60	10.70	3.36	1.56	0.79	0.44
700	–	–	–	37.70	14.50	4.55	2.13	1.09	0.59
800	–	–	–	49.00	19.00	5.89	2.77	1.42	0.78
900	–	–	–	62.30	24.10	7.60	3.51	1.80	0.99
1,000	–	–	–	76.90	29.80	9.30	4.35	2.21	1.22
1,500	–	–	–	–	67.00	21.00	9.80	4.90	2.73

[1] Compressed Air Handbook, Compressed Air and Gas Institute, 1947.

Table 9-4. Loss of Air Pressure Due to Friction in a Hose [1]
(in psi per 50 Feet of Hose With 100-Pound-Gauge Initial Pressure)

Cubic Feet of Free Air Per Minute	Nominal Hose Diameter (Inches)				
	1/2	3/4	1	1 1/4	1 1/2
20	0.6	0.2	–	–	–
30	2.0	0.4	0.1	–	–
40	4.3	0.6	0.2	–	–
50	7.6	1.0	0.2	–	–
60	12.0	1.4	0.4	–	–
70	17.6	2.0	0.5	0.1	–
80	24.6	2.7	0.6	0.2	–
90	33.3	3.5	0.8	0.2	–
100	44.5	4.4	1.0	0.3	–
110	–	5.4	1.2	0.4	–
120	–	6.6	1.5	0.4	0.1
130	–	7.9	1.8	0.5	0.2
140	–	9.4	2.1	0.6	0.2
150	–	11.1	2.4	0.7	0.2

[1] Compressed Air Handbook, Compressed Air and Gas Institute, 1947.

PAVING BREAKER (JACKHAMMER) (80-POUND)

9-20. The pneumatic paving breaker (80-pound weight class) *(Figure 9-3)* is used for heavy-duty demolition work on concrete, brick, asphalt, and macadam. It is also used for demolishing walls, columns, piers, and foundations and for general rock breaking. A variety of attachments may be used with this tool, depending on the type of work. This tool is a member of the reciprocating percussion family of air tools. It weighs 87.5 pounds, uses a 3/4-inch-diameter hose, and requires 65 cfm of air at 80 to 100 psi. It is constructed so that it may be separated into three major groups of parts: the back head, the cylinder, and the front head. The back-head group contains the air controls, the oil reservoir, and the handle. The cylinder group consists of the cylinder, the piston, and the automatic valve assembly. The front-head group provides the means for holding the tool steel or any attachment.

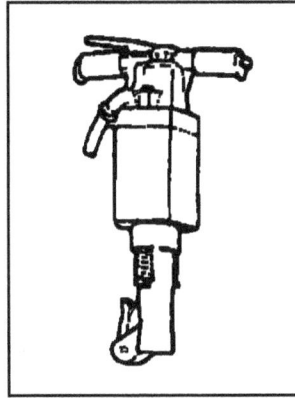

Figure 9-3. Paving Breaker (80-Pound)

Attachments

9-21. Four primary attachments are issued with this paving breaker. They are the moil point, the chisel point, the tamper, and the sheeting driver.

- The moil point is a 20-inch-long piece of 1 1/4-inch-diameter hexagonal tool steel that is pointed at one end and has a retainer collar 6 inches from the opposite end. It is used to break through concrete, stone, or other material having a similar high-abrasive and high-density character.
- The chisel point is similar to the moil point except that it has a 3-inch-wide working edge that is used to cut macadam, frozen ground, or extremely hard earth. It can be used for making a marking line to serve as a guide when cutting concrete.
- The tamper is a 5- to 7-inch-diameter steel pad, mounted on a 1 1/4-inch-diameter hexagonal tool steel. It is used to compact loose material.
- The sheeting driver is made of two steel angles and an impact pad that transmits the blow to the wood or metal sheeting that is being driven.

Production

9-22. Since job-site conditions and the mechanical condition of the air compressors and the pneumatic tools vary on each project, it is not possible to predict the work output of pneumatic tools on all jobs. In nonreinforced, 6- to 8-inch-deep concrete using a moil point, the average work output will range from 50 square feet per hour in large areas to 12 square feet per hour in narrow cuts. In reinforced concrete, production may drop to 50 square feet per 8-hour shift.

Operation

9-23. Hold down the paving breaker while it is in operation, but use only sufficient pressure to guide the tool and keep it in place. Leaning heavily on the paving breaker will shorten the stroke of the tool attachment and result in less work output. Breakers can best be operated in tandem. Only small bites (4 to 8 inches behind the working face) should be taken when breaking hard materials. If a moil point becomes stuck, use a second breaker to break the material binding the point. If the point becomes stuck when using a single breaker, take the paving breaker off and use another point to break the stuck point free. Other important operating precautions are as follows:

- Wear double hearing protection.
- Wear goggles to protect eyes from chips and dust.
- Ensure that the shank of each attachment is the correct size. Improper shank sizes will reduce the effectiveness of the blow and will cause damage to the paving breaker.
- Keep the points of the attachments sharp.
- Keep all nuts tight. The air hose to the paving-breaker connections should be checked frequently to ensure that no air is escaping.
- Provide a clear work area for efficient tool operation.

Maintenance

9-24. Maintenance problems inherent with the paving breaker are caused by improper use of the attachments. Too often, attempts are made to drill holes with the moil point. The moil point is a breaking device. Attempting to drill holes with it will break the point. The chisel point is designed for cutting asphalt and soft materials. If it is used for breaking concrete, the point will be damaged beyond repair. A frequent source of trouble with the paving breaker is breakage of the tool-latch retainer bolt. The cause of this is the operator not shutting off the tool when the moil point breaks through the material. The front head bounces on the concrete and causes the retainer bolt to break.

PAVING BREAKER (CLAY DIGGER) (25-POUND)

9-25. The pneumatic paving breaker (25-pound weight class) *(Figure 9-4, page 9-10)* is a medium-weight tool made for spading, trimming, cutting, or picking clay, hardpan, or frozen ground. It weighs 25.2 pounds, uses a 1/2-inch-diameter hose, and requires 35 cfm of air at 80 to 90 psi. It is constructed so that it may be separated into three major groups of parts: the back head, the cylinder, and the front head. The back-head group includes the handle. The cylinder group constitutes the main body of the tool. It includes the hammer, which is driven against the shank of the tool by the air pressure. The front-head group is the tool retainer.

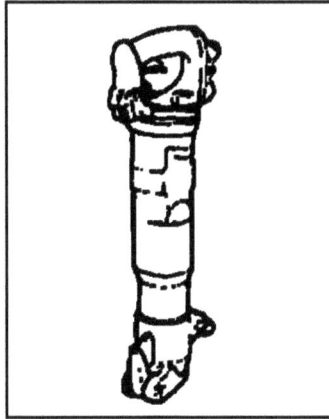

Figure 9-4. Paving Breaker (25-Pound)

Attachments

9-26. The three primary attachments normally issued with this breaker are the moil point, the pick, and the spade. A metal, drum-ripping tool may be issued for opening 55-gallon drums.

9-27. The moil point consists of a 15-inch straight length of 1-inch-diameter tool steel that is pointed on one end. It is used as a light demolition tool on masonry, concrete, or other dense material.

9-28. The pick's blade is 3 inches wide by 8 inches long with a pointed cutting end. It is used for digging into frozen ground, cemented gravel, or other materials too hard to be penetrated by the spade.

9-29. The spade (shaped like a garden spade) is 5 1/2-inches wide by 8-inches long. It is used for digging trenches, preparing footings or foundations, digging caissons, driving tunnels, or doing any general digging that is too difficult and slow for an ordinary hand spade.

9-30. The metal, drum-ripping tool has a 1-inch-wide cutting blade, topped by a 5/8-inch-thick, extended snubnose. The Army has the following two types—

- Type I is used to cut heads from metal drums. The nose of this ripping tool is curved to allow it to easily follow the curvature of the head on the drum.
- Type II is used to split metal drums lengthwise. It has a straight nose and is capable of opening 20 to 30, 55-gallon drums per hour.

Production

9-31. The attachment used most frequently with the 25-pound breaker is the clay spade. About 12 cubic yards of tough clay can be loosened per 10-hour shift.

Operation

9-32. Operators should merely guide the tool, never ride or lean on it. The tool is designed for trimming or digging, not for prying.

Maintenance

9-33. Give particular attention to the tool's retainer assembly. Dirt and other abrasive materials will enter the bottom of the retainer and cause excessive wear. This wear can be prevented if the tool is not allowed to penetrate past the wide portion of the clay spade.

NAIL DRIVER

9-34. The pneumatic nail driver *(Figure 9-5)* is a long-stroke, piston-action riveting hammer. The nail driver is designed for driving heavy drift pins and spikes. It weighs 24 pounds, uses a 1/2-inch-diameter hose, and requires 30 cfm of air at 90 psi. The handle is formed to fit the hand, with a thumb-operated throttle lever that controls the admission of air. The barrel of the driver houses the valve mechanism, the piston, and the nail set. A sleeve on the end of the nail set prevents the tool from sliding off the head of the nail. A safety set retainer screws onto the nozzle end of the barrel and holds the nail set in the tool at all times.

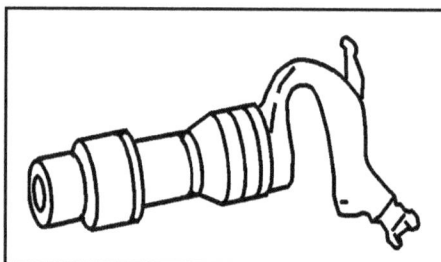

Figure 9-5. Nail Driver

Attachments

9-35. The nail driver is issued with 1/2- and 3/4-inch nail sets and a rivet buster.

Production

9-36. Used as a nail driver, 250 60-penny nails can be driven per hour (after the nails have been started by hand).

Operation

9-37. Always start the nails or spikes with a handheld hammer. The nail driver must be in line with the nail or spike being driven and should strike the nail or spike squarely to minimize the possibility of bending.

Maintenance

9-38. The retainer housing on a nail driver often breaks because the operator fails to keep the nail set against the work. Any attempt to countersink a nail will result in a broken retainer spring.

CIRCULAR SAW

9-39. The pneumatic circular saw *(Figure 9-6)* may be used for ripping and crosscut tasks. It weighs 32 1/2 pounds, uses a 1/2-inch-diameter hose, and requires 75 cfm of air at 80 to 100 psi. The handle assembly includes a D-shaped handle with a trigger-type throttle and a thumb-operated plunger lock. The body contains a rotary-vane air motor with a flyball governor that limits the motor speed to 2,400 rpm. A fixed blade guard is attached to the left side of the body to protect the operator. The top handle (above the body) is used to control and guide the saw. The foot is hinged to the front of the upper blade guard through a sector. By loosening a wing nut on this sector, the foot can be tilted for bevel cuts up to 45°. At the back of the foot a second sector, secured by a wing nut, permits adjustment of the depth of cut from 2 3/8 to 6 inches. Two V-shaped notches on the front of the foot simplify cutting along a line. The deeper V-notch is in line with the blade for right-angle cuts, while the smaller V-notch is in line with the blade for 45° bevel cuts. A rip fence (attached to the front of the foot by means of a wing screw) should be used for ripping when long cuts must be made. A telescopic blade guard covers the lower portion of the blade when the saw is not being used. This guard is spring-loaded so it closes automatically when the blade is lifted from the cut, but folds into the fixed blade guard when the saw is being operated.

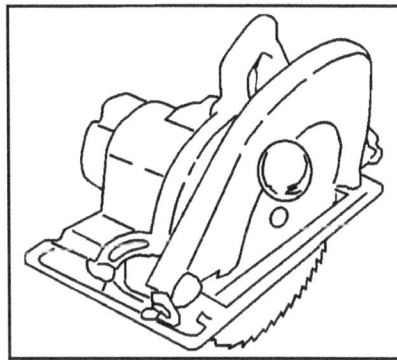

Figure 9-6. Circular Saw

Attachments

9-40. This saw is issued with a 12-inch combination blade used for ripping and crosscutting in wood only. When equipped with the proper abrasive disk, the pneumatic saw can be used to cut brick, stone, concrete, tile, asbestos cement sheets, steel, or cast iron. No one type of abrasive disk or saw blade is suited for all materials. Order these items carefully for each specific kind of work.

Production

9-41. Using the combination blade for crosscutting, the saw will cut a 4- by 4-inch board in 30 seconds. The maximum depth of cut at 90° is 4 3/8 inches.

Operation

9-42. Always use the proper blade for the material being cut. Make sure that the material to be cut is free of nails, spikes, or similar objects. Never jam the saw into a cut. If the saw is to be used upside down for prolonged periods of time, be careful that the exhaust port does not become clogged. Keep hands away from the blades, and shut off the air when the tool is not in use.

Maintenance

9-43. In many cases, the pneumatic circular saw is inverted and used as a table saw. When this is done, the exhaust port is exposed to the woodcuttings. An accumulation of these cuttings will clog up the air motor and make the saw useless.

CHAIN SAW

9-44. The pneumatic chain saw *(Figure 9-7)* is a heavy-duty saw intended primarily for cutting trees or timbers up to 24 inches in diameter. It weighs 45 pounds and requires 90 cfm of air at 80 to 100 psi. The hose diameter varies with the distance to the air source (25 feet or less from source, 5/8 inch; 26 to 100 feet from source, 3/4 inch; more than 100 feet from source, 1 inch). The head assembly consists of a drive housing, two handles, a guide bracket, a bumper, and an air connection. The drive housing contains a four-vane rotary motor. A guard bar made of heavy steel extends from the head assembly to the idler assembly and is slightly arched so it lies about 3/4 inch from the upper portion of the chain. Its purpose is to protect the operator from injury in the event of a break in the chain. The saw should never be operated without this guard. The guard bar issued with the chain saw is for the 24-inch-length saw; however, guard bars are available through supply channels for the 36- and 48-inch-length saws. Use of a 48-inch bar requires two operators.

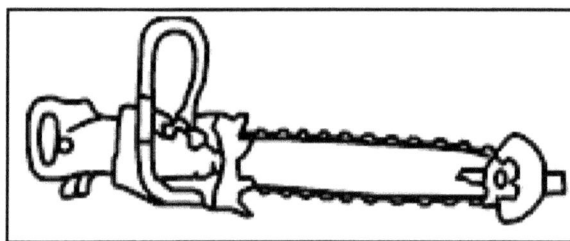

Figure 9-7. Chain Saw

Chains

9-45. The standard chain has a 3/4-inch pitch and a 3/8-inch cut for general-purpose use on any capacity saw. It is used for felling and for cutting hardwood or softwood. It is easy to sharpen and holds its cutting edge for a relatively long time. This chain consists of three-link sets. The link in the center of each set contains a raker tooth. Raker teeth are set alternately in the sets, to the right and left. The first and third links in each set contain a cutter tooth. The cutter teeth alternate on the chain, with the teeth set to the right and to the left. The cutter teeth control the width of the cut. A 76-inch chain is issued with the 24-inch-length chain saw; however, chains of 100 and 124 inches are available through supply channels for use with 36- and 48-inch-length saws.

Production

9-46. The chain saw can cut through a 12-inch hardwood log in 50 seconds. Never force the saw into the wood, but allow it to cut at its own speed. Be careful to ensure that the saw does not twist while cutting.

Maintenance

9-47. Keep the chain at the proper tension and properly sharpened. The blade should be adjusted to maintain a 1/2-inch chain slack when pulled up at the center. More slack than this will allow the chain to jump out of the saw guide, causing the blade to bend or break. If the chain is too tight, it will bend and cause sprocket damage.

WOOD DRILL

9-48. The pneumatic wood drill is a heavy-duty, low-speed tool designed to drive auger-type drill bits. It weighs 27 1/2 pounds, uses a 3/4-inch-diameter hose, and requires 60 cfm of air at 80 to 100 psi. It is used extensively in trestle bridge and other timber construction work where it is necessary to drill holes for bolts and pins. The drill body houses a rotary-vane air motor, a gear train (for reducing the motor speed to a chuck speed of about 800 rpm), and an oil reservoir. A chuck is provided for 1/2-inch-diameter drill-bit shanks and a large Allen-type setscrew holds the shank in place. There are two types of chucks—the Morse-taper and the two-screw. The shaft, on which the chuck is mounted, is drilled so the shank will extend into the base of the grip handle. A slot in the base of this handle provides for insertion of a wedge against the end of the bit to loosen it if it is jammed in the chuck. The air line is attached to the end of the throttle handle.

Attachments

9-49. Auger-type drill bits are issued in 1- and 3-foot lengths and have 7/16-, 3/4-, 1-, and 2-inch diameters.

Production

9-50. The drill will bore 125 36-inch-deep holes in one hour using a 2-inch-diameter auger bit.

Operation

9-51. The rotation of the wood drill can be reversed. Always start the drill slowly until the screw is well set. Hold the drill firmly, but do not force it. Exert enough effort to counteract the tendency of the tool to rotate, and be prepared to resist the torque in case the bit becomes stuck. During boring and withdrawing of the auger, keep it in line with the hole.

Maintenance

9-52. The auger bit frequently becomes stuck in the chuck. Remove it by using the auger ejector. Trying to knock it out with a hammer will result in damage to the chuck and/or the auger.

SUMP PUMP

9-53. The pneumatic sump pump *(Figure 9-8)* is a small-capacity pump. The sump pump weighs 50 pounds, uses a 3/4-inch-diameter hose, and requires 100 cfm of air at 80 to 90 psi. Due to its simple, rugged construction it requires little attention. It can operate while completely submerged when an exhaust line is used. The pump assembly consists of an open-impeller centrifugal pump. A combination bottom plate and inlet strainer cover the pump intake opening, and a 3-inch exhaust connection is mounted on the side of the pump housing.

Figure 9-8. Sump Pump

Production

9-54. The pneumatic sump pump may be either a Class 1 (for transferring sewage and sludge) or a Class 2 (for transferring petroleum products). This pump is rated at 175 gallons per minute (GPM) against a 25-foot head or 150 GPM against a 150-foot head.

Operation

9-55. To ensure maximum efficiency, keep the inlet strainer clean and free of debris. Keep the pump away from mud, and clean the strainer as often as is necessary. Keep the exhaust-line outlet above the water level. Use only water-pump grease in the fittings on the pump. Drain the pump of water when not using it.

Maintenance

9-56. If silt and dirt are left in the pump after use, it will cause the impeller to stick and will require disassembly and cleaning before it can be used again. Allowing water to get into the pump through the exhaust port will cause failure of the grease seals.

STEEL DRILL

9-57. The pneumatic steel drill is a portable tool for drilling, reaming, and tapping in metals. The drill weighs 27.5 pounds, uses a 1/2-inch-diameter hose, and requires 27 cfm of air at 90 to 100 psi. The chuck speed is 425 rpm. It is suitable for 1 1/4-inch drilling and 1-inch reaming or tapping.

Attachments

9-58. Bits for use with this drill are 1/2-inch in diameter with a Number 3 Morse-taper shank.

Production

9-59. Used as a drill, thirty 1-inch holes can be drilled per hour if the steel plate has been prepared beforehand with 1/4-inch lead holes.

Operation

9-60. The rotation of the steel drill cannot be reversed. It is important to ensure that the bits have clean, sharp edges, and that they are not chipped or damaged in any way.

- Use cutting oil to cool and lubricate the drill bit.
- Use a center punch to mark the center and to hold the tip of the drill in place when starting a hole.
- Do not use worn chucks.
- Wear goggles to protect eyes from steel chips or shavings.
- Clamp all material that is being drilled to a bench. This will prevent injuries to personnel if the drill should bind in the material.

Maintenance

9-61. The bit will be damaged due to heat if cutting oil is not used. Too much pressure applied to the bit will stall the drill and cause undue wear on the gear assembly. This can damage the feed-screw system.

HANDHELD ROCK DRILL

9-62. The pneumatic handheld rock drill is a piston-action unit with independent air-motor rotation. It is designed primarily as a hard-rock drill; however, it is also efficient in soft and medium formations. It weighs 57 pounds, uses a 3/4-inch-diameter hose, and requires 95 cfm of air at 80 to 100 psi. The drill consists of a back-head group, a cylinder unit, and a front-head group. It is designed so that air may be directed through the drill, down the drill steel, and into the bottom of the hole to blow out rock cuttings.

Attachments

9-63. This drill is issued with drill rods in 2-, 4-, 6-, and 8-foot lengths and drill-bits of 1 5/8, 1 3/4, 1 7/8, and 2 inches.

Production

9-64. The drill is designed for vertical drilling. If large numbers of horizontal holes are required, some mechanical means must be devised for holding the drill in place. It will drill holes efficiently to a depth of 10 feet. See *Table 9-2, page 9-3,* for production rates.

Operation

9-65. Bent drill steels should not be used. They cause damage to the drill and usually result in a stuck bit and lost production.

SAFETY

9-66. Be very careful when working with compressed air. At close range it is capable of putting out eyes, bursting eardrums, causing serious skin blisters, or even killing an individual.

AIR COMPRESSORS

- Ensure that the intake air is cool and free from flammable gases or vapors.
- Do not permit wood or other flammable materials to remain in contact with the air-discharge pipe.
- Shut down the compressor immediately if the air discharged from any stage rises unduly or exceeds 400°F.
- Ensure that all the pressure gauges are in good working order.
- Do not kink a hose to stop the air flow.
- Check the safety valves, pressure valves, and regulators to determine if they are working properly before starting the air compressor.
- Do not leave the compressor after starting it, unless you are certain that the control, unloading, and governing devices are working properly.
- Do not run an air compressor faster than the manufacturer's recommended speed.
- Use only the proper grade of oil as recommended by the manufacturer.
- Use only oils which have high flash points to lubricate the air cylinders.
- Avoid the application of too much oil.
- Keep the compressor, the tanks, and the accompanying piping clean to guard against oil-vapor explosion.
- Clean the intake air filters periodically.
- Use only soapy water or a suitable nontoxic, nonflammable solution for cleaning compressor intake filters, cylinders, or air passages. Never use benzene, kerosene, or other light oils to clean these portions of a system. These oils vaporize easily and the vapor will explode when compressed.
- Turn off the motor before making adjustments and repairs.
- Make certain that the compressor is secured and cannot be started automatically or by accident, that the air pressure in the compressor is completely relieved, and that all the valves between the compressor and the receivers are closed before working on or removing any part of the compressor.

PNEUMATIC TOOLS

- Wear protective clothing and equipment (such as goggles, gloves, and respirators) appropriate for the particular pneumatic tool being operated.
- Maintain a firm grip on the tool at all times.
- Maintain a good footing and proper balance at all times while operating pneumatic tools.

- Release the throttle of the tool at the first indication that the tool is out of control. Release the tool and let it fall if it cannot be controlled.
- Turn off the air and disconnect the tool when repairs or adjustments are being made or the tool is not in use. When disconnecting the tool, all pressurized air should be discharged from the line before the connection is broken.
- Inspect the hose to ensure that it is in good condition and free from obstructions before connecting a pneumatic tool. When blowing out the line, make certain the end of the hose is pointed into the air and is secured against whipping. Make certain all connections are tight before the line is pressurized.
- Lay down pneumatic tools in such a manner that no harm can be done if the switch is accidentally tripped. Do not leave an idle tool in a *standing* position.
- Keep pneumatic tools in good operating condition and thoroughly inspect them at regular intervals. Give particular attention to the control and exhaust valves, the hose connections, the guide clips on hammers, and the chucks of reamers and drills.
- Shut off the tool and relieve the pressure from the line before disconnecting the tool from the line.
- Remove leaking or defective hoses from service. The air hose must be suitable to withstand the pressure required for the tool.
- Do not lay the hose over ladders, steps, or walkways in such a manner as to create a tripping hazard.
- Where a hose is run through a doorway, protect the hose against damage from the door's edge.
- Do not lay the hose between the operator's legs while the tool is being operated.
- Never point an air hose at other personnel. Do not use compressed air to clean clothing being worn or to blow dust off the body.

Handheld Rock Drills

- Do not (under any circumstances) wear loose or torn clothing.
- Examine the drill for defects. Pay particular attention to bit flutes, which must be ground to uniform size, sharpness, and length.
- Hold the machine on a straight line with the hole being bored.
- Do not feed the machine too fast.
- Establish a firm footing before starting the operation.
- Do not modify or bypass the handgrip switch. (All drills are equipped with a handgrip switch that will shut off the air supply when the grip is released.)

Paving Breakers

- Wear suitable goggles when operating pneumatic breakers.
- Roughen hard materials or slick surfaces with a sledgehammer to improve breaker control.

Chapter 10

Hauling Equipment

The most common hauling equipment used for Army construction work are the 5- and 20-ton dump trucks, both of which are organic to most engineer units. Equipment trailers are used to transport heavy construction equipment not designed for cross-country travel. They are also used to haul long, oversize items and packaged items.

DUMP TRUCKS

USE

10-1. The 5-ton family of medium tactical vehicles (FMTV) *(Figure 10-1)* and the 20-ton *(Figure 10-2, page 10-2)* dump trucks can be used for a variety of purposes. This manual, however, discusses dump trucks used primarily for hauling, dumping, and spreading earth, rock, or processed aggregates.

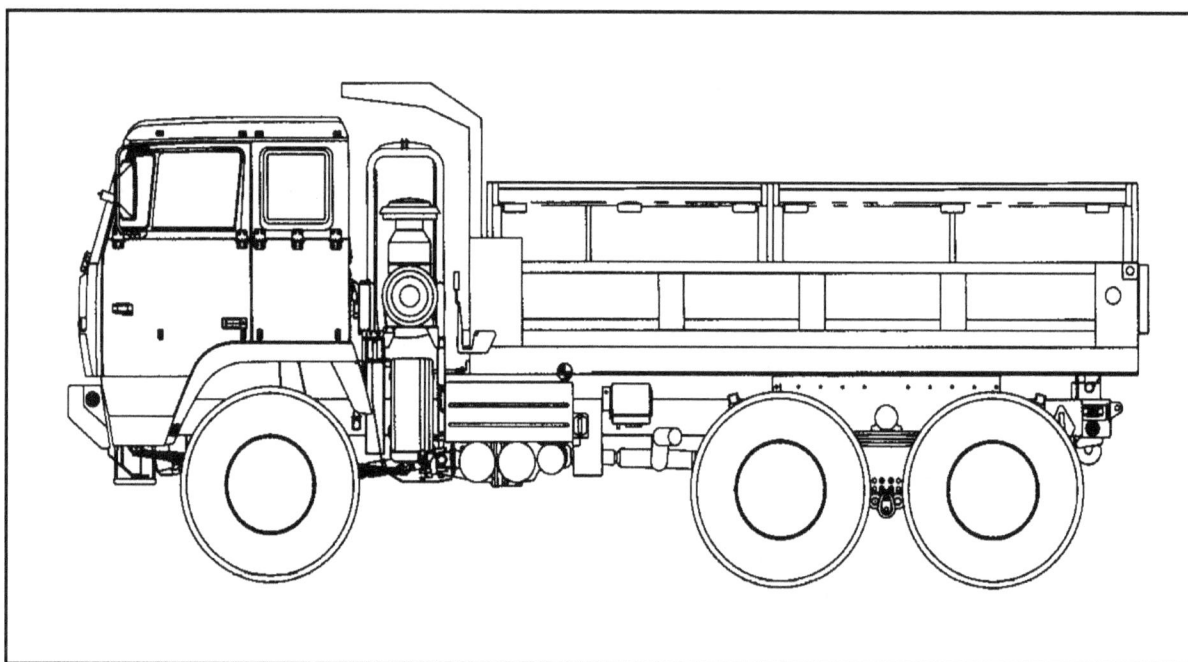

Figure 10-1. Dump Truck (5-Ton) FMTV

CAPACITY

10-2. The capacity of hauling equipment is expressed in one of three ways: gravimetrically by the weight of the load it will carry (in tons), by its struck rear-dump body volume (in cubic yards), or by its heaped rear-dump body capacity (in cubic yards). The hauling capacity of Army dump trucks is normally expressed gravimetrically: 5-ton and 20-ton. Conversely, the capacity of loading equipment is normally expressed in cubic yards. The unit weight of the various materials to be transported may vary from as little as 1,700 pounds per LCY for dry clay, to 3,500 pounds per LCY for trap rock (see *Table 1-2, page 1-4,* for weights of common materials). Always make sure that the volumetric load does not exceed the gravimetric capacity of the truck.

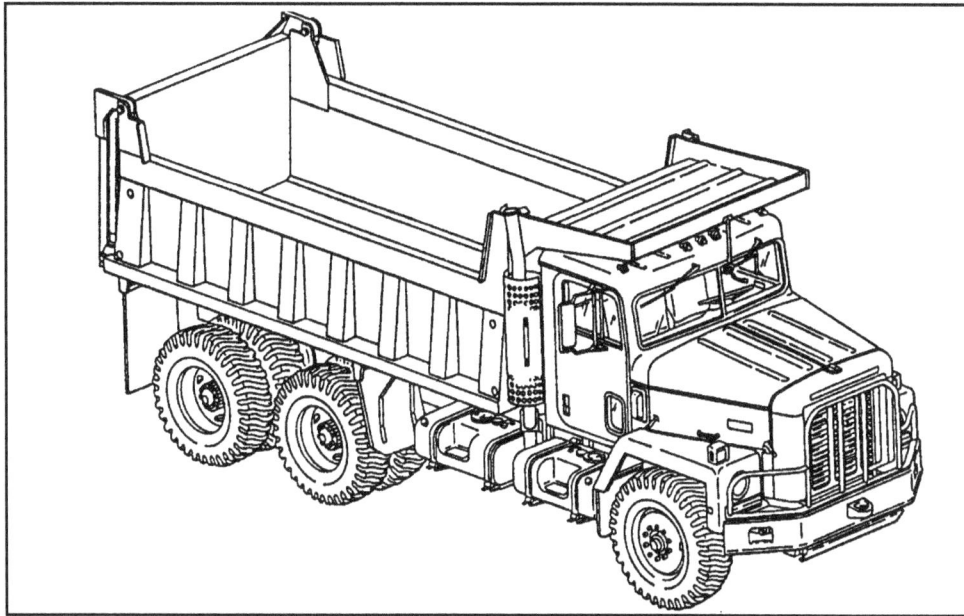

Figure 10-2. Dump Truck (20-Ton)

OPERATION
Loading

10-3. For maximum efficiency, fill trucks as close to their rated hauling capacity as practical. Adjust the load size if haul roads are in poor condition or if the trucks must traverse steep grades. Overloading will cause higher fuel consumption, reduced tire life, and increased mechanical failures.

10-4. Use spotting markers when trucks are hauling from a hopper, a grizzly ramp, or a stockpile. Spotting markers are also beneficial when excavators (such as a dragline, a clamshell, a loader, a backhoe, or a hoe) are used to load hauling equipment. They facilitate prompt and accurate vehicle spotting which improves loading efficiency.

10-5. Spot trucks as close to the bank as possible when loading with an excavator. Ensure that the trucks are within the working radius of the dragline, the clamshell, or the hoe bucket. When using a loader, position the truck and loader so that the two machines form a V. This arrangement will reduce the loader cycle time*(Figure 10-3).*

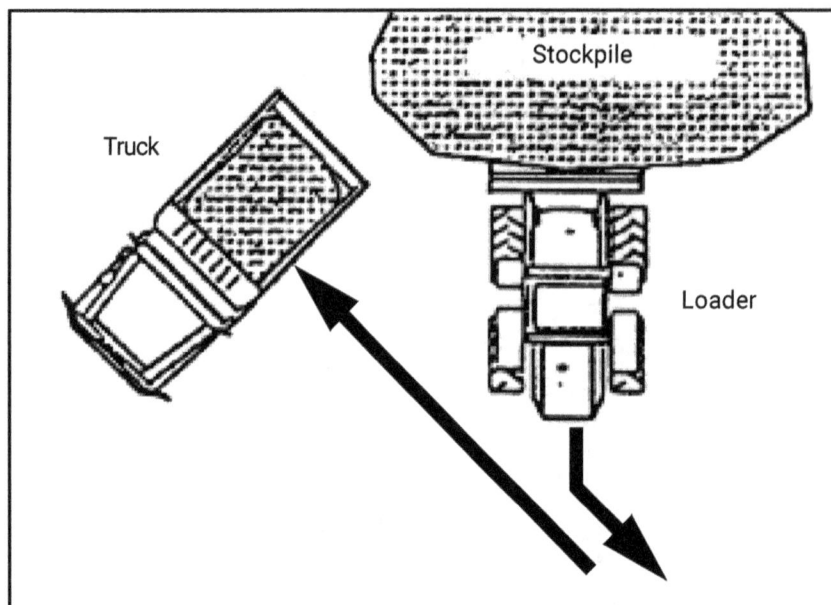

Figure 10-3. Truck and Loader V-Positioning for Loading

Maintaining Proper Speed

10-6. Haul at the highest safe speed and in the proper gear, without speeding. Speeding is unsafe and hard on equipment. When several trucks are hauling, it is essential to maintain the proper speed to prevent hauling delays or bottlenecks at the loading and dumping sites. Use separate haul roads to and from the dump site, if possible. Keep haul roads well maintained, with a minimum grade. Use one-way traffic patterns to increase efficiency.

Dumping (Unloading)

10-7. Always use spotters to control dumping operations. When dumping material that requires spreading, move the truck forward slowly while dumping the load. This makes spreading easier. Establish alternative dumping locations to maintain truck spacing when poor footing or difficult spotting slow the dumping operation.

Preventive Maintenance

10-8. Keep truck bodies clean and in good condition. Accumulations of rust, dirt, dried concrete, or bituminous materials hamper production. Consider the time spent cleaning and oiling dump bodies, particularly for asphalt or concrete hauling, when computing transportation requirements.

- Clean truck bodies thoroughly at the end of the day. When used to haul wet concrete mix, spray the dump beds with water before loading and clean them thoroughly as soon as practical after dumping.
- Coat the walls and sides of truck bodies with diesel fuel or oil to prevent bituminous materials (plant-mix asphalt) from sticking.

PRODUCTION ESTIMATES

10-9. The production capacity of the loading equipment is normally the hauling operation's controlling factor. Never keep loading equipment waiting. If there are not enough trucks, there will be a loss in production.

Number of Trucks Required

10-10. Use the following formula to estimate the number of trucks required to keep loading equipment operating at maximum capacity:

$$\text{Number of trucks required} = 1 + \frac{\text{truck cycle time (minutes)}}{\text{loader cycle time (minutes)}}$$

- The numeral 1 in the formula is a safety factor against the necessity for closing down loading equipment due to lack of hauling equipment. If all operations are on schedule, one truck will always be standing by at the loader, ready for spotting.
- The truck cycle time is the time required for a truck to complete one cycle of operation. One complete cycle is the time a loaded truck takes to travel to the dump site, unload, return to the loading unit, and be reloaded.
- The loader cycle time is the time it takes the loading equipment to load the truck, plus any time lost by the loading equipment while waiting for the truck to be spotted.

NOTE: After the job has started, the number of trucks required may vary because of changes in haul road conditions, reductions or increases in haul length, or changes in conditions at either the loading or unloading areas.

Number of Standby Trucks Required

10-11. Identify, based on the normal cycle time, the number of standby trucks that should be available to replace trucks that develop mechanical trouble. The number of standby trucks needed depends largely on the mechanical condition of the active trucks as well as the size and importance of the job. In the case of a small fleet and a single loading unit, the ratio of standby trucks to active trucks may be as high as 1:4. On larger jobs, the ratio is smaller. Standby trucks need not be idle; use them on lower priority tasks from which they can easily be diverted.

EXAMPLE

How many 5-ton FMTV trucks (hauling 3 LCY per load) will it take to support a wheel loader having a 2-cubic-yard heaped-bucket capacity? The haul-unit cycle time is 20 minutes excluding loading time. The loader cycle time per bucket load is 0.5 minute. Consider a 60-minute working hour.

Step 1. Determine the number of bucket loads required to fill a truck.

$$\text{Bucket loads} = \frac{\text{haul-unit capacity}}{\text{bucket capacity}} = \frac{3 \text{ LCY}}{2 \text{ LCY}} = 1.5 \text{ bucket loads}$$

Using only one bucket load would mean that the truck would only haul 2 LCY per trip. Using two bucket loads would mean that the truck would haul 4 LCY per trip and the extra material would spill out during the loading process.

Step 2. Determine the loading time per haul unit.

Loading time per haul unit = bucket cycle time x number of bucket loads

Considering one bucket load per truck—

Loading time per haul unit = 0.5 minute x 1 = 0.5 minute

Considering two bucket loads per truck—

Loading time per haul unit = 0.5 minute x 2 = 1 minute

Step 3. Determine the number of hauling units needed to support the loading unit.

Considering one bucket load per truck—

Truck cycle time = 20 minutes + 0.5 minute = 20.5 minutes

$$\text{Number of trucks required} = 1 + \frac{\text{truck cycle time (minutes)}}{\text{loader cycle time (minutes)}} = 1 + \frac{20.5 \text{ minutes}}{0.5 \text{ minute}} = 42 \text{ trucks}$$

Considering two bucket loads per truck—

Truck cycle time = 20 minutes + 1 minute = 21 minutes

$$\text{Number of trucks required} = 1 + \frac{\text{truck cycle time (minutes)}}{\text{loader cycle time (minutes)}} = 1 + \frac{21 \text{ minutes}}{1 \text{ minute}} = 22 \text{ trucks}$$

Step 4. Determine the production based on the number of hauling units used.

The loader will control the production because of the one extra truck added to the formula. Therefore, there is always a truck waiting at the loader.

$$\text{Production} = \text{haul-unit load} \times \frac{\text{minutes per working hour}}{\text{loader cycle time in minutes}}$$

Using one bucket load per truck will require 42, 5-ton FMTV dump trucks.

$$\text{Production} = 2 \text{ LCY} \times \frac{60}{0.5} = 240 \text{ LCY per hour}$$

Using two bucket loads per truck will require 22, 5-ton FMTV dump trucks.

$$\text{Production} = 3 \text{ LCY} \times \frac{60}{1} = 180 \text{ LCY per hour}$$

With an understanding of the effect of the different choices, determine the number of trucks to use on the haul and how many bucket loads to place on each truck. This illustrates that the capacity of both the loader and the trucks are set numbers. Therefore, there is a relationship between bucket loads and haul-unit capacity, which in practice must be an integer number.

EQUIPMENT TRAILERS

USE

10-12. Use equipment trailers *(Figure 10-4)* to transport heavy construction equipment such as cranes, dozers, or any equipment not designed for long-distance movement by their own power. Also use the trailers to haul long items such as pipes or lumber, or packaged items such as landing mats or bagged cement.

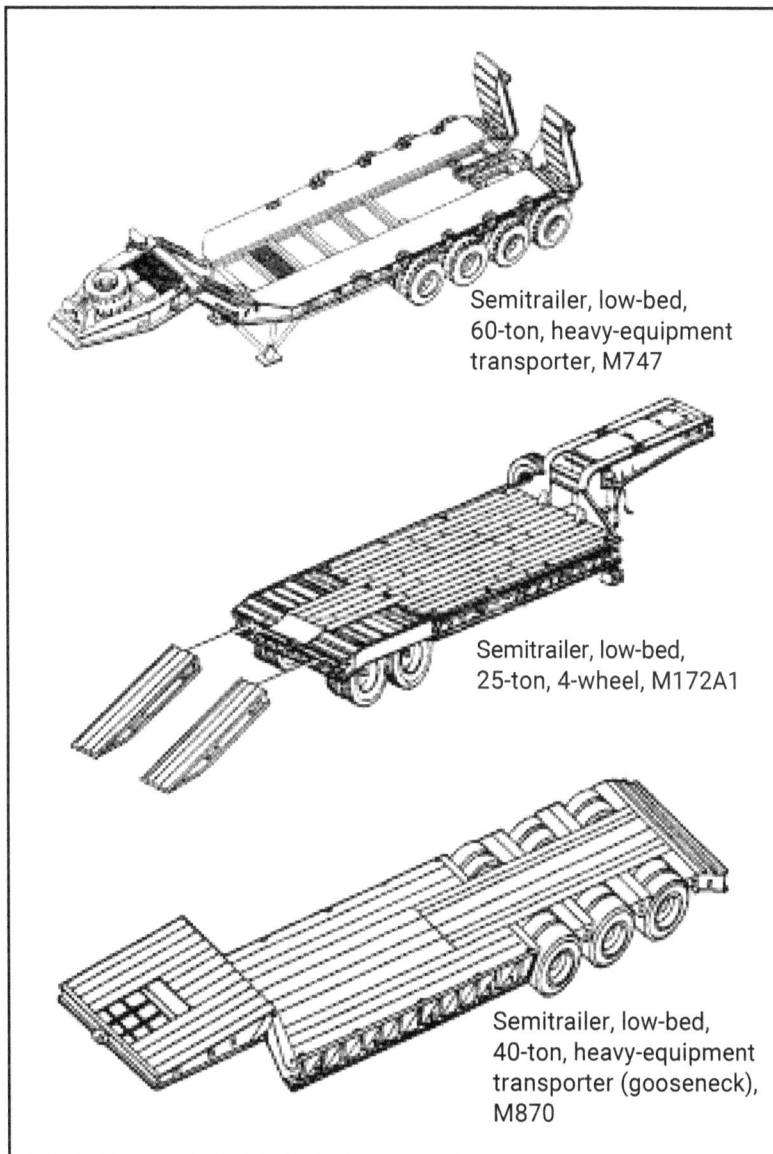

Semitrailer, low-bed, 60-ton, heavy-equipment transporter, M747

Semitrailer, low-bed, 25-ton, 4-wheel, M172A1

Semitrailer, low-bed, 40-ton, heavy-equipment transporter (gooseneck), M870

Figure 10-4. Equipment Trailers

OPERATION
Loading

10-13. For maximum efficiency, load trailers as close as possible to their rated loading capacity. When loading, always station a spotter on the trailer to direct the equipment operator and to keep the machine centered on the ramps and trailer.

10-14. With rear-loading trailers, use low banks or built-up earth ramps where possible. Some trailers carry loading ramps for loading from level ground. When using loading ramps to load a dozer, run the machine slowly up the ramps (with the blade raised) and as the balance point is reached, reduce speed or stop, then lower the blade and allow the front of the tracks to settle gently onto the trailer bed. Then move the dozer slowly ahead onto the trailer. Some low-bed trailers are designed for front-end loading.

10-15. In areas that restrict rear loading, load the trailer from the side. Take care not to damage the trailer bed.

NOTE: Refer to the unit's SOP or to the appropriate technical manual for proper techniques for loading and securing equipment.

Positioning and Securing

10-16. After positioning the equipment on the trailer bed, block and chock it and chain it to the trailer. Properly distribute the weight of large equipment on the trailer. Trailers have their load-weight centering position marked.

Unloading

10-17. Unload heavy equipment slowly to prevent damage to the trailer or the equipment. Always use ramps to load and unload.

Chapter 11

Soil-Processing and Compaction

Horizontal construction projects such as roads and airfields are constructed using a variety of soil types. The suitability of these materials for construction applications depends on their gradation, physical characteristics, and load-bearing capacity. While some soil types are suitable for structural purposes in their natural state, others require processing such as adjusting the moisture content by mixing and blending. Because there is a direct relationship between increased density and increased strength and bearing capacity, the engineering properties of most soils can be improved simply by compaction. Soil properties and compaction requirements are discussed in FM 5-410.

SOIL PROCESSING

11-1. The amount of water present in a soil mass affects the ease of compaction operations and the achievable soil density. The water-content ratio is the standard measure of water in a soil mass. The water-content ratio compares the weight of the water present in a soil mass to the weight of the soil solids in the same mass. Each soil has its particular optimum moisture content (OMC) at which a corresponding maximum density can be obtained for a given amount of compactive input energy. Trying to compact a soil at a water content either higher or lower than optimum can be very difficult. The OMC varies from about 12 to 25 percent for fine-grained soils and from 7 to 12 percent for well-graded granular soils. Since it is difficult to attain and maintain the exact OMC, normal practice is to work within an acceptable moisture range. This range, which is usually ±2 percent of optimum, is based on attaining the maximum density with the minimum compactive effort. Determination of the OMC is a laboratory test procedure. For a detailed description of the moisture-density relationships of various soils, refer to FM 5-410.

INCREASING THE MOISTURE CONTENT

11-2. If the moisture content of a soil is below its optimum moisture range, add water to the soil before compaction. When it is necessary to add water, the project officer must consider the following:

- The amount of water required.
- The rate of water application.
- The method of application.
- The effects of the weather.

Add water to the soil at the borrow pit or in place (at the construction site). When processing granular materials, adding water in place usually gives the best results. After adding water, thoroughly and uniformly mix it with the soil.

Amount of Water Required

11-3. It is essential to determine the amount of water required to achieve a soil water content within the acceptable moisture range. Compute the amount of water to add or remove in gallons per station (100 feet of length). Use the following formula, based on the compacted volume, to compute the amount of water to add or remove from the soil. The volume in this formula is for only one station of project length. The computation is based on the dry weight of the soil.

Gallons per station for one lift = desired dry density of soil in pounds per cubic foot (pcf)

$\times \dfrac{\text{desired moisture content (percent)} - \text{moisture content of borrow (percent)}}{100}$

$\times \dfrac{\text{compacted volume of soil (cubic foot)}}{8.33 \text{ pounds per gallon}}$

where—

8.33 = the weight of a gallon of water

NOTE: Normally, it is a good practice to adjust the desired moisture content to OMC +2 percent, but this depends on the environmental conditions (temperature and wind) and the soil type. A negative answer indicates that water removal from the borrow material is necessary before compacting the material on the fill.

EXAMPLE

Prepare to place soil in 6-inch (compacted) lifts. The desired dry unit weight of the embankment is 120 pcf. The OMC (desired moisture content) of the soil is 12 percent, but the soils technician has determined that the moisture content of the borrow material is only 5 percent. The roadway section to be placed is 40 feet wide. Compute the amount of water (in gallons) to add per station for each lift of material.

Gallons per station for one lift $= 120 \text{ pcf} \times \dfrac{12 \text{ percent (OMC)} - 5 \text{ percent (borrow)}}{100}$

$\times \dfrac{40 \text{ feet} \times 100 \text{ feet} \times 0.5 \text{ foot}}{8.33 \text{ pounds per gallon}}$

$= 120 \text{ pcf} \times 0.07 \times \dfrac{2{,}000 \text{ cubic feet}}{8.33 \text{ pounds per gallon}}$

$= 2{,}017$

NOTE: If the road width is constant, determine the total amount of water required for the job by multiplying the gallons per lift times the number of lifts, times the road length (in stations).

Rate of Water Application

11-4. After determining the total amount of water required, determine the rate of application. Use the following formal to determine the water application rate in gallons per square yard.

$$\text{Gallons per square yard} = \text{desired dry density of soil (pcf)}$$
$$\times \frac{\text{percent of moisture added or removed}}{100} \times \text{lift thickness (feet)}$$
$$\times \frac{\text{9 square feet per square yard}}{\text{8.33 pounds per gallon}}$$

where—
9 = factor used to convert square feet to square yards
8.33 = the weight of a gallon of water

EXAMPLE

Using the data from the previous example, determine the required application rate in gallons per square yard.

$$\text{Gallons per square yard} = 120 \text{ pcf} \times 0.07 \times 0.5 \text{ foot} \times \frac{\text{9 square feet per square yard}}{\text{8.33 pounds per gallon}}$$

= 4.5 gallons per square yard

Method of Application

11-5. After calculating the application rate, determine the method of application. Regardless of the method of application, it is important to achieve the proper application rate and the uniform distribution of water.

11-6. **Water Distributor.** The most common method of adding water is with a water distributor. Water distributors are designed to distribute the correct amount of water evenly over the fill. The truck-mounted, 1,000-gallon water distributor *(Figure 11-1, page 11-4)* can distribute water under various pressures or by gravity feed. It distributes the water through a 12-foot folding, rear-mounted spray bar. The spray bar is adjustable, in 1-foot increments, from 4 to 24 feet. The water application rate can be maintained by controlling the forward speed of the vehicle and the water distribution pressure. A cab-mounted odometer shows the vehicle speed in fpm. The project officer should provide the water-distributor operator with the application rate in gallons per square yard. With this information, the operator can determine the appropriate spray-bar length, pumping pressure, and vehicle speed to achieve the required application rate. Refer to the vehicle's technical manual for specific information regarding application rates.

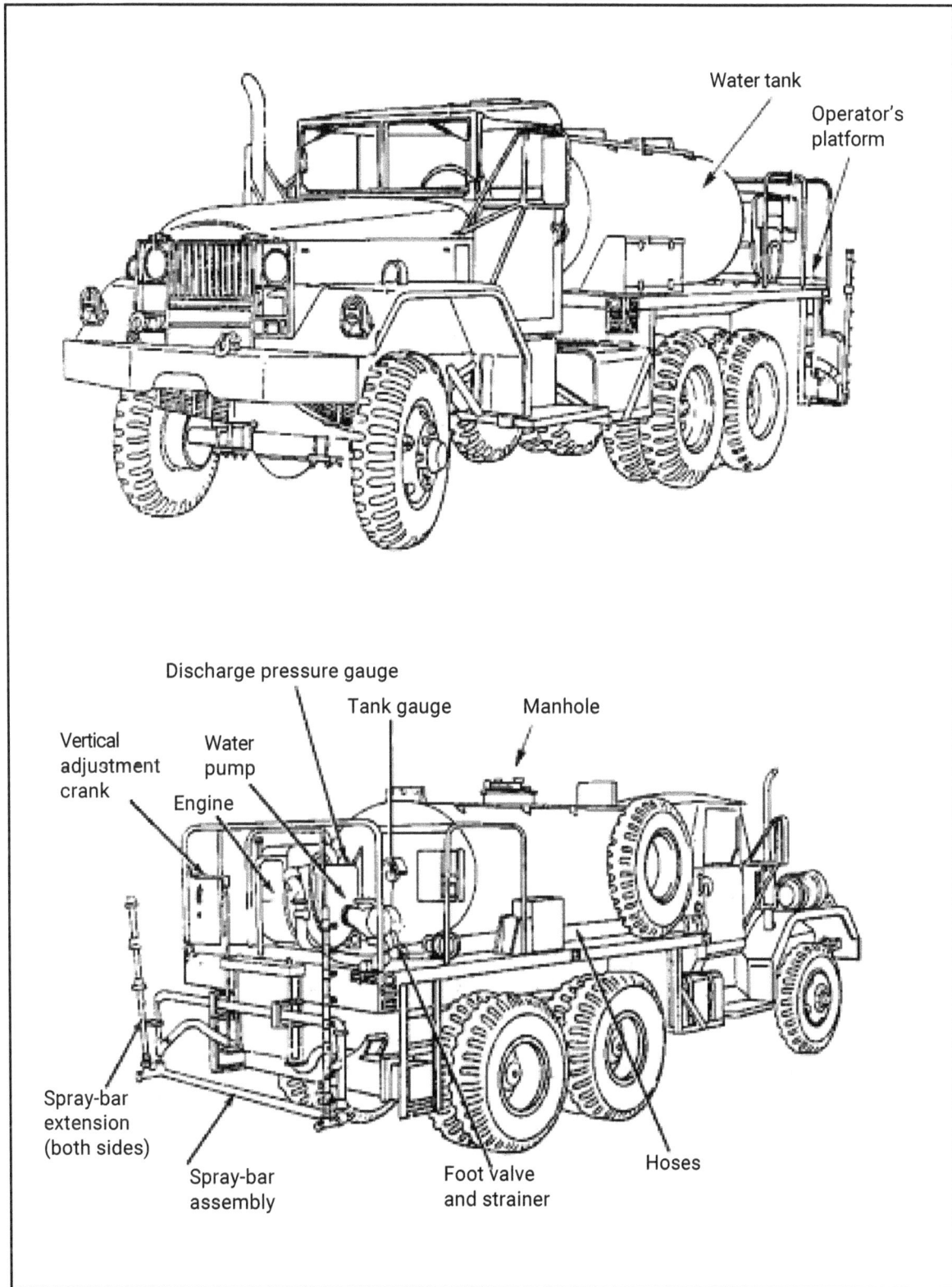

Figure 11-1. Truck-Mounted, 1,000-Gallon Water Distributor

11-7. Ponding. If time is available, add water by ponding the area until achieving the desired depth of penetration. It is difficult to control the application rate with this method. Ponding usually requires several days to achieve a uniform moisture distribution.

Effects of the Weather

11-8. Weather substantially affects the soil's moisture content. Cold, rainy, cloudy, or calm weather will cause a soil to retain water or even increase its moisture content. Hot, dry, sunny, or windy weather is conducive to drying the soil by evaporating the moisture. In a desert climate, evaporation claims a large amount of water intended for the soil lift. Thus, for a desert project the engineer might go as high as 6 percent above the OMC as a target for all water application calculations. This allows the actual moisture content to fall very near to the desired content when placing and compacting the material.

REDUCING THE MOISTURE CONTENT

11-9. As previously stated, soil that contains more water than desired (above the optimum moisture range) is correspondingly difficult to compact. Excess water makes achieving the desired density very difficult. In these cases, take action to reduce the moisture content to within the required moisture range. Drying actions may be as simple as aerating the soil. However, they may be as complicated as adding a soil stabilization agent that changes the physical properties of the soil. Lime or fly ash are the typical stabilization agents for fine-grained soils. Excess moisture, caused by a high water table, will require some form of subsurface drainage to reduce the soil's moisture content. The most common method of reducing the moisture is to scarify the soil prior to compaction. Accomplish this by using the scarifying teeth on a grader or a stabilizer mixer or by disking the soil. Another method is to use the grader's blade to toe the soil over into furrows to expose more material for drying.

MIXING AND BLENDING

11-10. Whether adding water to increase the soil's moisture content or adding a drying agent to reduce it, it is essential to mix the water or drying agent thoroughly and uniformly with the soil. Even if additional water is not necessary, mixing may still be essential for a uniform distribution of the existing moisture. Accomplish mixing by using graders, stabilizer mixers, or farm disks.

Grader

11-11. Use conventional graders to mix or blend a soil additive (water or stabilizing agent) by windrowing the material from one side of the working lane to the other. For a detailed description of grader operation, refer to *Chapter 4.*

Stabilizer Mixer

11-12. The stabilizer mixer is an extremely versatile piece of equipment designed specifically for mixing, blending, and aerating materials *(Figure 11-2, page 11-6).* The stabilizer consists of a rear-mounted, removable-tine, rotating tiller blade covered by a removable hood. In place, the hood creates an enclosed mixing chamber, which enhances thorough blending of the soil *(Figure 11-3, page 11-6).* The tiller blade lifts the material in the direction of travel and throws it against the leading edge of the hood. The material deflecting off of the hood falls back onto the tiller blades for thorough blending. As the stabilizer moves forward, it ejects the material from the rear of the mixing chamber. As the material is ejected, it is struck off by the trailing edge of the hood, resulting in a fairly level working surface. With the trailing edge of the hood fully opened, churned soil has a very high void content, which exposes the soil to the drying action of the sun and wind. Models equipped with a spray bar are used to add water or stabilizing agents to the soil during the blending process. The stabilizer mixer's use is limited to material less than 4 inches in diameter. The tines on the Army's mixer are designed to penetrate up to 12 inches below the existing surface. This unit is used for scarifying and blending in-place (in situ) material as well as fill material.

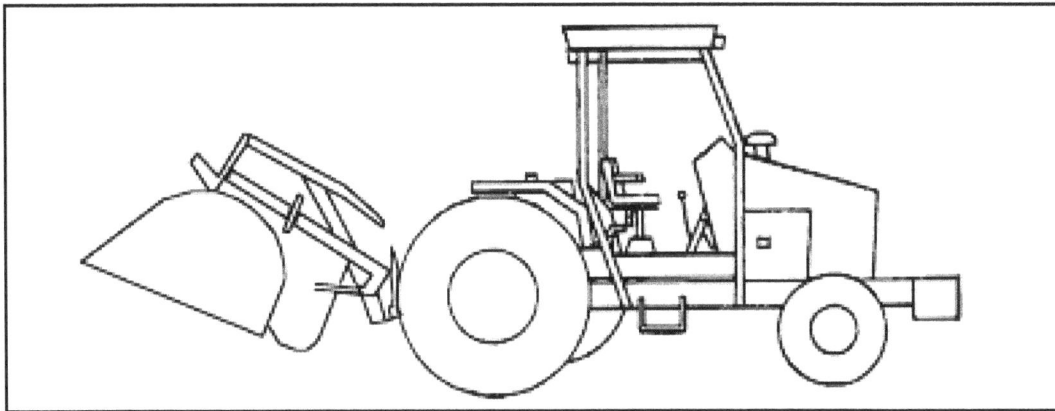

Figure 11-2. Stabilizer Mixer

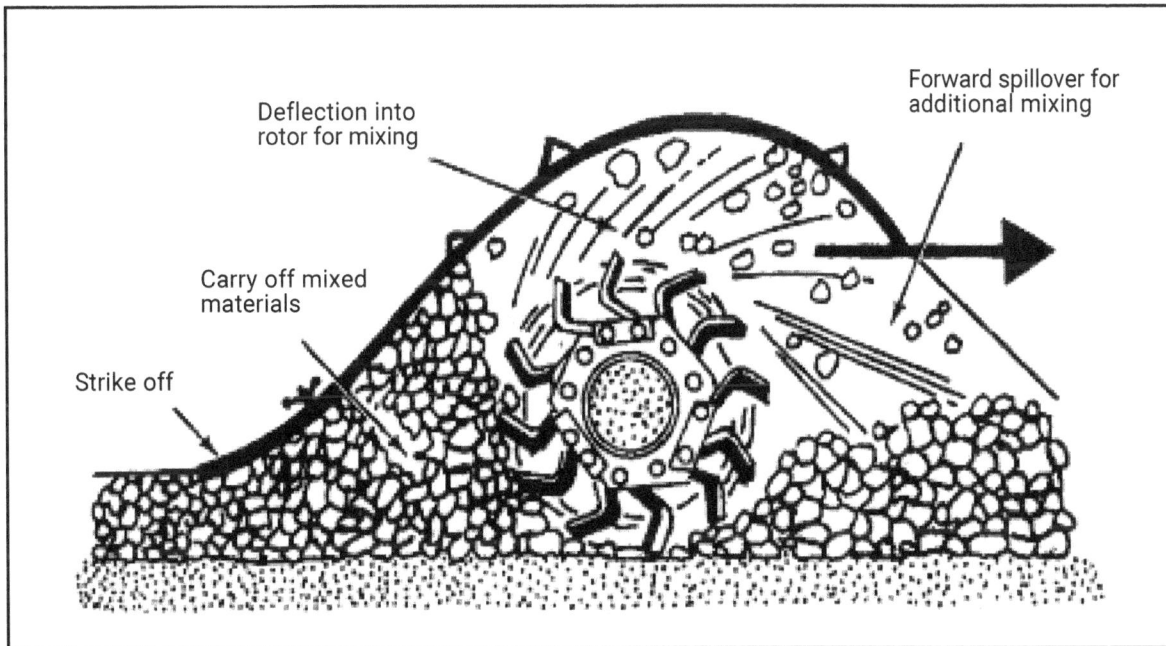

Figure 11-3. Mixing Action in a Stabilizer Mixer

SOIL COMPACTION

11-13. Compaction is the process of mechanically densifying a soil, normally by the application of a moving (or dynamic) load. This is in contrast to consolidation, which is the gradual densification of a soil under a static load. When controlled properly, compaction increases a soil's load-bearing capacity (shear resistance), minimizes settlement (consolidation), changes the soil's volume, and reduces the water-flow rate (permeability) through the soil. Compaction does not affect all soils to the same degree. However, the advantages gained by compaction make it an essential component of the horizontal construction process.

COMPACTIVE EFFORT

11-14. Compactive effort is the amount of energy used to compact a soil mass. Base the appropriate compactive effort on the physical properties of the soil, including gradation (well or poorly graded), the Atterberg limits (cohesive or cohesionless), and the required final density. Compaction equipment uses one or more of the following methods to accomplish soil densification—

- Static weight (pressure).
- Kneading (manipulation).
- Impact (sharp blow).
- Vibration (shaking).

EQUIPMENT SELECTION

11-15. Compaction equipment ranges from handheld vibratory tampers (suitable for small or confined areas) to large, self-propelled rollers and high-speed compactors (ideally suited for large, horizontal construction projects). Consider the following factors when selecting compaction equipment:

- Type and properties of the soil.
- Density desired.
- Placement lift thickness.
- Size of the job.
- Compaction equipment available.

11-16. Soil-compacting equipment normally available to military engineers includes tamping-foot rollers, pneumatic-tired (rubber-tired) rollers, dual-drum vibratory rollers, and smooth-drum vibratory rollers. To select the most appropriate type of compaction equipment, a project officer must know the characteristics, capabilities, and limitations of the different types of rollers. Generally, tamping-foot compactors that produce high unit pressures are best for predominantly fine-grained cohesive materials such as clays and sandy clays. Large, steel-drum rollers are best for larger particle materials such as gravel or cobble. Vibratory rollers are ideal for well-graded or gap-graded materials because the shaking action causes the smaller particles to fill voids around the larger grains. *Table 11-1, page 11-8,* shows the spectrum of capabilities for each type of roller and the type of compactive effort associated with each roller. *Tables 11-2* and *11-3, pages 11-9* and *11-10,* show the major soil-classification categories, the compaction requirements, and the compactive methods compatible with each.

Table 11-1. Compaction-Equipment Capabilities

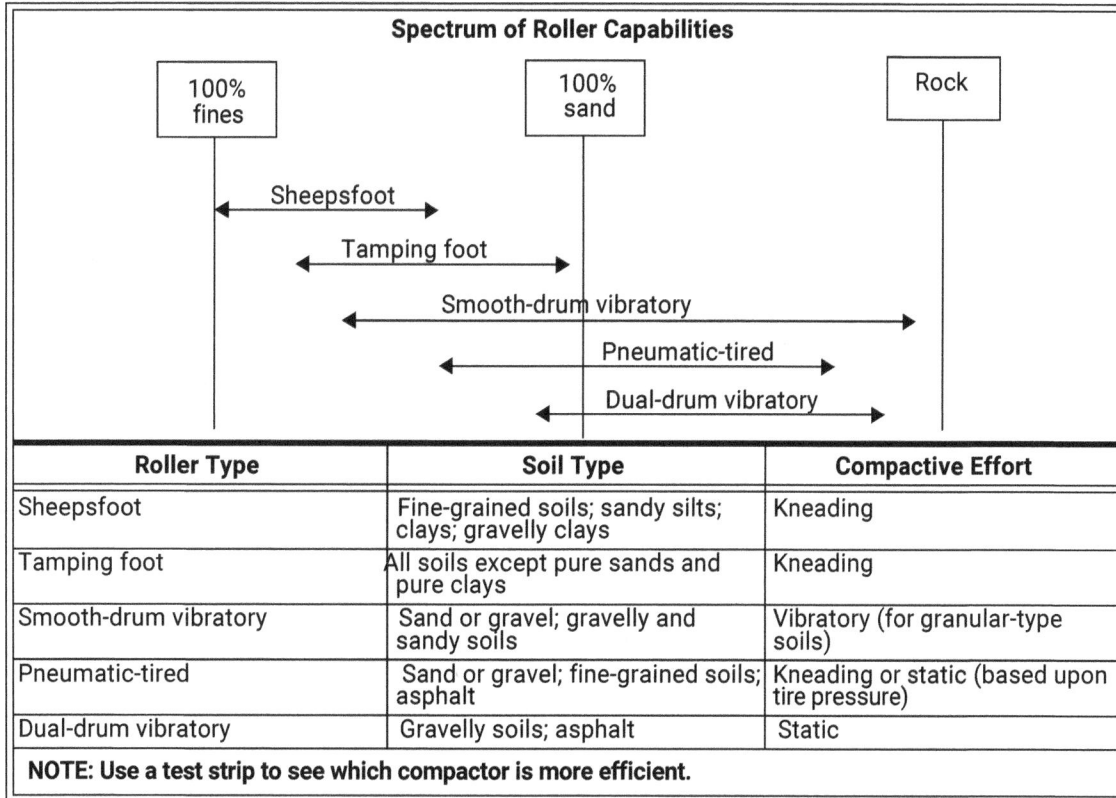

Spectrum of Roller Capabilities		
100% fines	100% sand	Rock

Sheepsfoot

Tamping foot

Smooth-drum vibratory

Pneumatic-tired

Dual-drum vibratory

Roller Type	Soil Type	Compactive Effort
Sheepsfoot	Fine-grained soils; sandy silts; clays; gravelly clays	Kneading
Tamping foot	All soils except pure sands and pure clays	Kneading
Smooth-drum vibratory	Sand or gravel; gravelly and sandy soils	Vibratory (for granular-type soils)
Pneumatic-tired	Sand or gravel; fine-grained soils; asphalt	Kneading or static (based upon tire pressure)
Dual-drum vibratory	Gravelly soils; asphalt	Static
NOTE: Use a test strip to see which compactor is more efficient.		

Tamping-Foot Roller

11-17. The self-propelled, tamping-foot roller *(Figure 11-4, page 11-11)* has feet that are square or angular and taper down away from the drum. This design allows the roller to achieve better penetration on the initial pass, resulting in a thorough, uniform compaction throughout a lift. This roller compacts the material from the bottom of the lift to the top, and walks out after achieving the desired density. It is suitable for compacting all fined-grained materials, but is generally not suitable for use on cohesionless granular materials. The lift thickness for the tamping-foot roller is limited to 8 inches in compacted depth. If the material is loose and reasonably workable (permitting the roller's feet to penetrate into the layer on the initial pass), it is possible to obtain a uniform density throughout the full depth of the lift. Thoroughly loosen material that has become compacted by the wheels of equipment during spreading or wetting before compaction. The tamping-foot roller does not adequately compact the upper 2 to 3 inches of a lift. Therefore, follow up with a pneumatic-tired or smooth-drum roller to complete the compaction or to seal the surface if not placing a succeeding lift. The self-propelled tamping-foot roller can achieve a working speed of as high as 8 mph. The tamping-foot roller compacts from the bottom up and is particularly appropriate for plastic materials. It is ideal for working soils that have moisture contents above the acceptable moisture range since it tends to aerate the soil as it compacts.

Table 11-2. Soil Classification

Major Soil Categories		Symbol and Description		Value as a Base, Subbase, or Subgrade	Potential Frost Action
Coarse-grained soils (50% or more larger than a #200 sieve opening)	Gravel and/or gravelly soils	GW	Well-graded gravels or gravel-sand mixture with 5% or less of fines	Fair to good for base; good to excellent for subbase and subgrade	None to very slight
		GP	Poorly graded gravels or gravel-sand mixture with little or no fines	Fair to good for all	None to very slight
		GM	Silty gravel and poorly graded gravel-sand-silt mixtures	Not suitable for base (15% or less of fines with PI of 5 or less); fair to excellent for subbase and subgrade (50% or less of fines)	Slight to medium
		GC	Clayey gravel and poorly graded gravel-sand-clay mixture	Not suitable for base (15% or less of fines with PI of 5 or less); poor to good for subbase and subgrade	Slight to medium
	Sand and/or sandy soils	SW	Well-graded sands or gravelly sand mixture with 5% or less of fines	Poor for base; fair to good for subbase and subgrade	None to very slight
		SP	Poorly graded sands or gravelly sand mixture with 5% or less of fines	Poor to not suitable for base; poor to fair for subbase and subgrade	None to very slight
		SM	Silty sands, sand-silt mixture	Not suitable for base; poor to good for subbase and subgrade	Slight to high
		SC	Clayey sands, sand-clay mixture	Not suitable for base; poor to fair for subbase and subgrade	Slight to high
Fine-grained soils (more than 50% smaller than a #200 sieve opening)	Silt and clays with liquid limits less than 50	ML	Inorganic silt of low plasticity, silty fine sands	Not suitable for base or subbase; poor to fair for subgrade	Medium to very high
		CL	Inorganic clay of low to medium plasticity, lean clays	Not suitable for base or subbase; poor to fair for subgrade	Medium to high
		OL	Organic silt and organic silt-clay of low plasticity	Not suitable for base or subbase; poor to very poor for subgrade	Medium to high
	Silt and clays with liquid limits greater than 50	MH	Inorganic silt micaceous or diatomaceous soil	Not suitable for base or subbase; poor to fair for subgrade	Medium to very high
		CH	Inorganic clay of high plasticity, fatty clays	Not suitable for base or subbase; poor to fair for subgrade	Medium
		OH	Organic clay of medium to high plasticity	Not suitable for base or subbase; poor to very poor for subgrade	Medium
Highly organic soils		Highly organic soils (peat) are not defined by numerical criteria; these soils are identified by visual and manual inspection.			

Table 11-3. Average Compaction Requirements

		Soil Classification Symbol													
		GW	GP	GM	GC	SW	SP	SM	SC	ML	CL	OL	MH	CH	OH
Sheepsfoot, Standard With Ballast (Towed by Dozer)	Lift Thickness Compacted (Inches)	*	*	*	6	*	*	*	*(Best)	6	6	6(Best)	6	6	(Best) 6
	Rolling Speed (mph)	NA	NA	NA	3	NA	NA	NA		NA	3		3		2 2 2 2
	Number of Passes	NA	NA	NA	10	NA	NA	NA	10	10	12	12	12	4	14
Self-Propelled Vibratory Roller	Lift Thickness Compacted (Inches)	18 (Best)	18 (Best)	12		12 (Best)	18 (Best)	18	2		12		8		8 *
	Rolling Speed (mph/vpm)	4/1,400 or more	4/1,400 or more	4/1,100	4/700 to none	4/1,400 or more	4/1,400 or more	4/1,100	3/700 to none	3/700 to none	3/700 to none	NA	NA	NA	NA
	Number of Passes	8	8	6	6	8	8	6	7	7	7	NA	NA	NA	NA
Tamping-Foot Roller, Self-Propelled (Not Recommended for Finishing Grade)	Lift Thickness Compacted (Inches)	12		12		9		9		12			12		9 9
	Rolling Speed (mph)	10	10	10	8	10	10	10		10		8	8		4 4 4
	Number of Passes	5	5	6	7	5	5	6	6	5	5		5	6	6 6
13-Wheel Pneumatic Compactor with Ballast (Wheel Towed), 100 psi	Lift Thickness Compacted (Inches)	6	6	6	6	6	6	6	6	4	4	4	4	4	4 4
	Rolling Speed (mph)	5	5	4	4	5	5	4	3	3	3	3	3		2 2
	Number of Passes	10	10	10	10	10	10	10	12	7	7	7	8	9	9
9-Wheel Pneumatic, Self-Propelled with Ballast, 100 psi	Lift Thickness Compacted (Inches)	6	6	6	6	6	6	6	6	4	4	4	4	4	4 4
	Rolling Speed (mph)	6	6	6	5	6	6	6	5	4	4	4	4	3	3
	Number of Passes	6	6	7	7	7	7	8	8	6	6	6	6	6	6
Smooth-Drum Vibratory Roller	Lift Thickness Compacted (Inches)	12		12		9		9		12			12		9 9
	Rolling Speed (mph/vpm)	4/1,400 or more	4/1,400 or more	4/1,100	4/700 to none	4/1,400 or more	4/1,400 or more	4/1,100	3/700 to none	3/700 to none	3/700 to none	NA	NA	NA	NA
	Number of Passes	8	8	8	9	8	8	8		0	1	0	1	0	N A

NOTES:
This chart should be used as a planning guide when a test strip cannot be performed.
The above symbols are based on the United Soil Classification System (USCS).
*Not recommended.

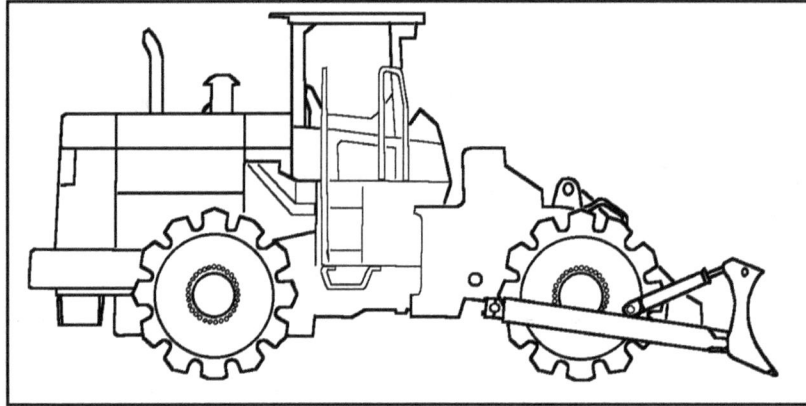

Figure 11-4. Self-Propelled, Tamping-Foot Roller

Pneumatic-Tired Roller

11-18. Pneumatic-tired rollers (towed and self-propelled) are suitable for compacting most granular materials. They are not effective in compacting fine-grained clays. Pneumatic-tired rollers compact using two types of compactive effort—static-load and kneading. The Army currently has a towed, 13-wheel, pneumatic-tired roller *(Figure 11-5)* and a variable-pressure, self-propelled, nine-wheel, pneumatic-tired roller *(Figure 11-6, page 11-12).* The nine-wheeled model is capable of varying the contact pressure to achieve the desired compactive effort. The contact pressure is controlled by adjusting the tire pressure and the wheel load. The towed, 13-wheel model exerts about 210 pounds of contact pressure per inch of rolling width. Contact pressure is affected by tire pressure and wheel load.

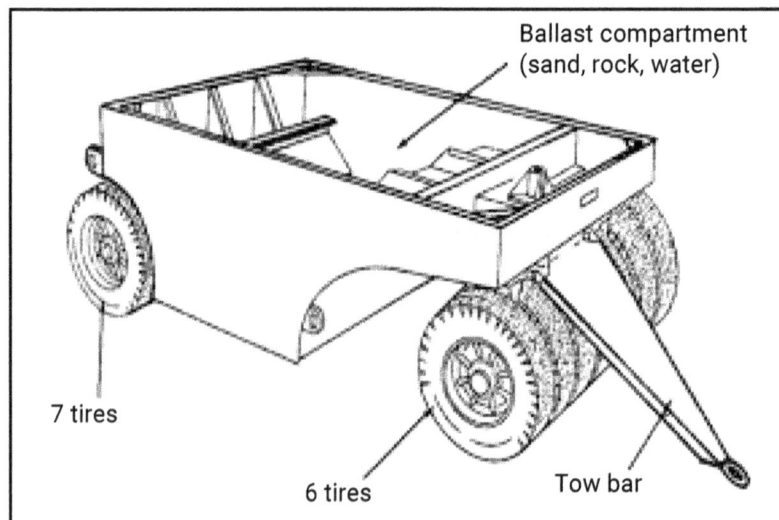

Figure 11-5. Towed, 13-Wheel, Pneumatic-Tired Roller

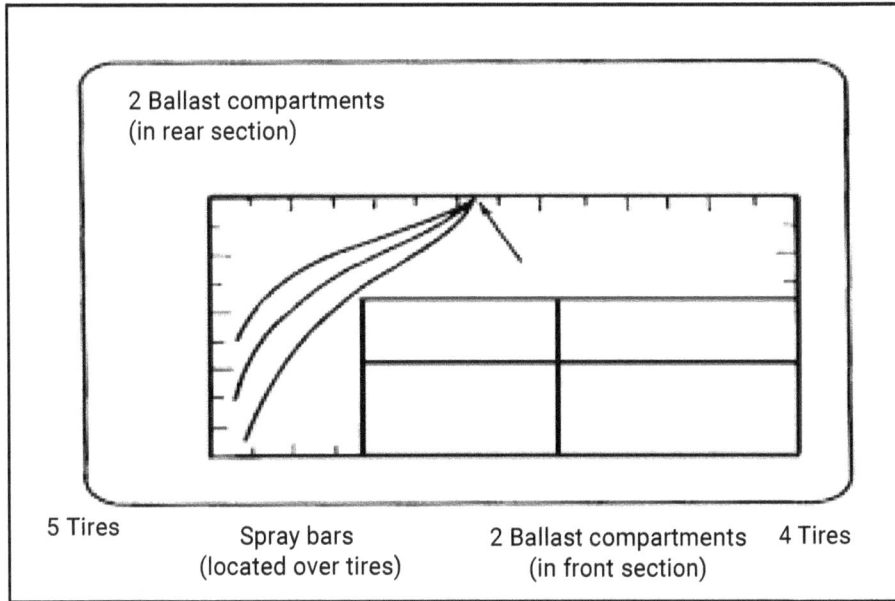

Figure 11-6. Self-Propelled, Nine-Wheel, Pneumatic-Tired Roller

- **Contact pressure.** The contact pressure of these rollers is determined primarily by the tire pressure. Within the rated load limits, the same load and tire pressure give about the same contact area for any tire. The tire sidewalls carry about 10 percent of the load, and the trapped air essentially supports 90 percent of the load. Consequently, the tire will deflect until the contact area is adequate and the ground pressure on the tire is equal to the tire pressure. For example, the contact area for a tire with a 50-psi internal tire pressure and a 5,000-pound wheel load is 100 square inches. If the wheel load is doubled to 10,000 pounds, the tire will deflect until 200 square inches are in contact with the ground. Since the sidewalls carry 10 percent of the load, the contact area is—

$$\text{Contact area} = \frac{0.9 \times \text{wheel load}}{\text{tire pressure}}.$$

Generally, the analysis of contact pressure neglects the raised portions of the tread. Use the gross contact areas, including the areas between the raised portion, to determine contact pressure.

$$\text{Contact pressure} = \frac{\text{wheel load}}{\text{contact area}}.$$

- **Wheel load.** The wheel load is significant for compacting at the required depth or in test rolling to detect subsurface defects. Researchers have built test sections in 6-inch compacted layers with wheel loads of 10,000; 20,000; and 40,000 pounds to determine if increased wheel loads would increase density. In the tests, the tire's inflation pressure was maintained at a constant 65 psi. *Figure 11-7* shows the vertical pressure distribution for the tire loadings. As shown, the effective pressure varies with the depth. However, at shallow depths, the pressure difference among the three loads was not enough to produce additional density. These and other tests have indicated that an increase in wheel load is advantageous in compacting thick lifts.

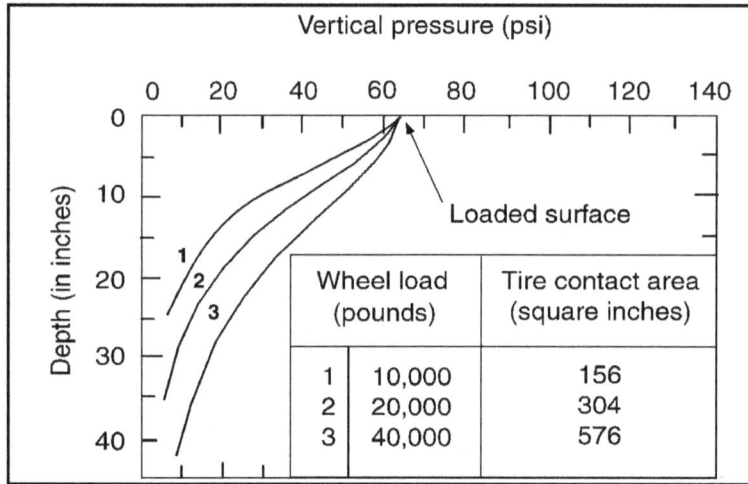

Figure 11-7. Vertical Pressure Distribution Beneath a Wheel Load

- **S u r f a c** The wheel arrangement and the tire deflection determine the surface coverage. *Figure 11-8* shows the results of varying wheel loads and tire pressures on single-pass coverage for a heavy pneumatic-tired roller. Most of the pneumatic-tired rollers use two rows of tires. The tires of one row offset the gaps between the tires of the second row. This ensures complete coverage with one pass. Heavier rollers have only one row of tires and require two passes for complete surface coverage. The additive effects of the pressure bulbs from the wheels on heavier rollers affect the at-depth coverage and the rolling pattern. *Figure 11-9, page 11-14,* shows that at-depth coverage requires considerable overlap with each pass to ensure that the entire area has received the same compactive effort.

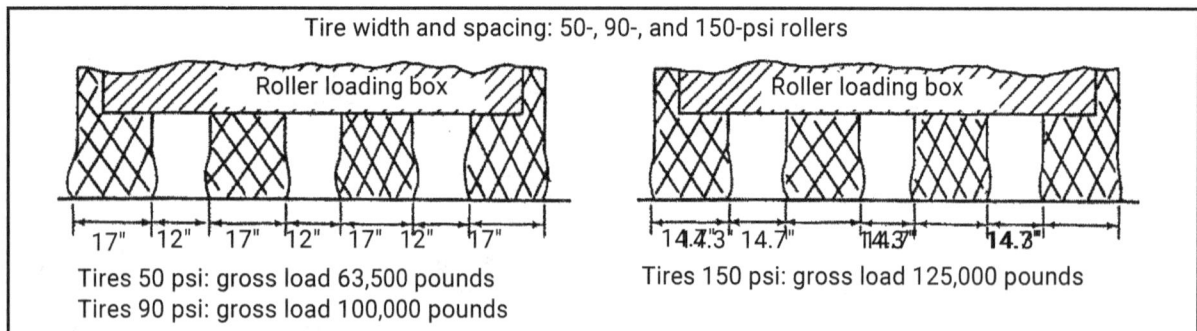

Figure 11-8. Varying Wheel Loads and Tire Pressures

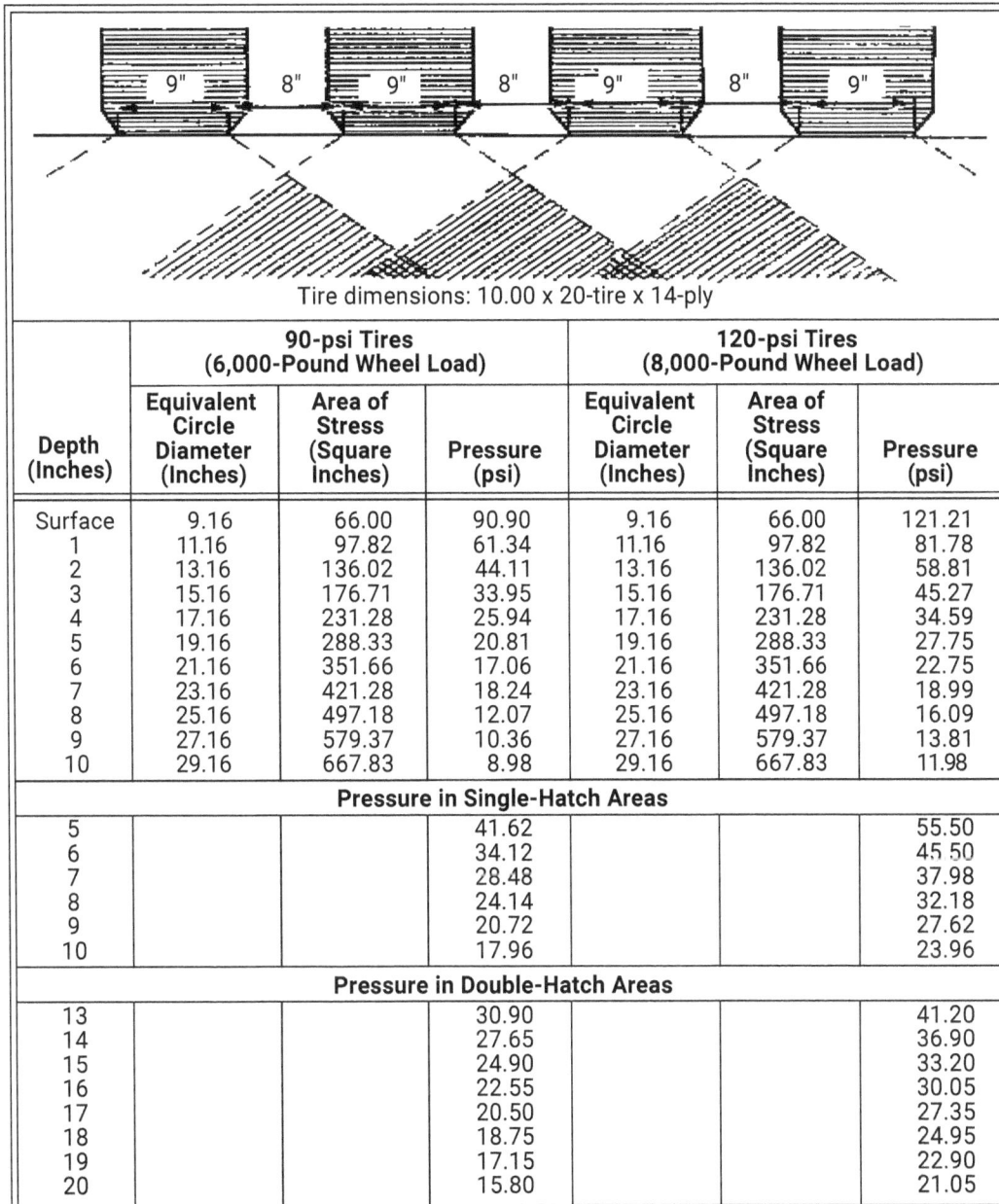

Tire dimensions: 10.00 x 20-tire x 14-ply

Depth (Inches)	90-psi Tires (6,000-Pound Wheel Load)			120-psi Tires (8,000-Pound Wheel Load)		
	Equivalent Circle Diameter (Inches)	Area of Stress (Square Inches)	Pressure (psi)	Equivalent Circle Diameter (Inches)	Area of Stress (Square Inches)	Pressure (psi)
Surface	9.16	66.00	90.90	9.16	66.00	121.21
1	11.16	97.82	61.34	11.16	97.82	81.78
2	13.16	136.02	44.11	13.16	136.02	58.81
3	15.16	176.71	33.95	15.16	176.71	45.27
4	17.16	231.28	25.94	17.16	231.28	34.59
5	19.16	288.33	20.81	19.16	288.33	27.75
6	21.16	351.66	17.06	21.16	351.66	22.75
7	23.16	421.28	18.24	23.16	421.28	18.99
8	25.16	497.18	12.07	25.16	497.18	16.09
9	27.16	579.37	10.36	27.16	579.37	13.81
10	29.16	667.83	8.98	29.16	667.83	11.98
Pressure in Single-Hatch Areas						
5			41.62			55.50
6			34.12			45.50
7			28.48			37.98
8			24.14			32.18
9			20.72			27.62
10			17.96			23.96
Pressure in Double-Hatch Areas						
13			30.90			41.20
14			27.65			36.90
15			24.90			33.20
16			22.55			30.05
17			20.50			27.35
18			18.75			24.95
19			17.15			22.90
20			15.80			21.05

Figure 11-9. Tire Pressures at Various Depths

Dual-Drum Vibratory Roller

11-19. The dual-drum vibratory roller, with its smooth steel drum *(Figure 11-10),* can compact a wide variety of materials from sand to cobble. Use this roller to compact asphalt paving, cohesionless subgrade, base course, and wearing surfaces. It is very effective when used immediately behind a blade to create a smooth, dense, and watertight subgrade finish. Because it has a relatively low unit pressure and compacts from the top down, it is normally used for relatively shallow lifts (less than 4 inches). The smooth steel drum is ideal for base- or wearing-course finish work. Exercise care to prevent excess crushing of the base-course material. This is the roller of choice for asphalt paving.

Figure 11-10. Dual-drum vibratory roller

Smooth-Drum Vibratory Roller

11-20. The smooth-drum vibratory roller *(Figure 11-11, page 11-16)* uses a vibratory action in conjunction with the ballast weight of the drum to rearrange the soil particles into a dense soil mass. Vibratory compaction, when properly controlled, can be one of the most effective and economical means of attaining the desired density for cohesionless materials. This roller is very effective in compacting noncohesive/nonplastic sands and gravels, which are often used in subbase and base-course applications. Because this roller is relatively light, the recommended maximum loose-lift depth (the fill material measured before compaction) is 9 inches.

NOTE: Vibration has two measurements—amplitude (the measurement of the movement or throw) and frequency (the number of repetitions per unit of time). The amplitude controls the depth to which the vibration is transmitted into the soil, and the frequency determines the number of blows or oscillations that are transmitted in a period of time.

Jay Tamper

11-21. The jay tamper is a small, self-contained, hand-operated, vibratory compactor. The jay tamper *(Figure 11-12, page 11-16)* resembles a power lawn mower. The jay tamper is ideal for compacting materials in confined spaces such as in a trench against a pipe or inside an existing structure. A gasoline engine powers the unit, so there are no supply lines or other auxiliary items to hinder its operations.

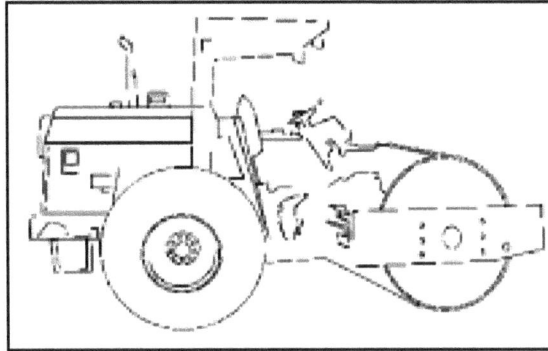

Figure 11-11. Smooth-Drum Vibratory Roller

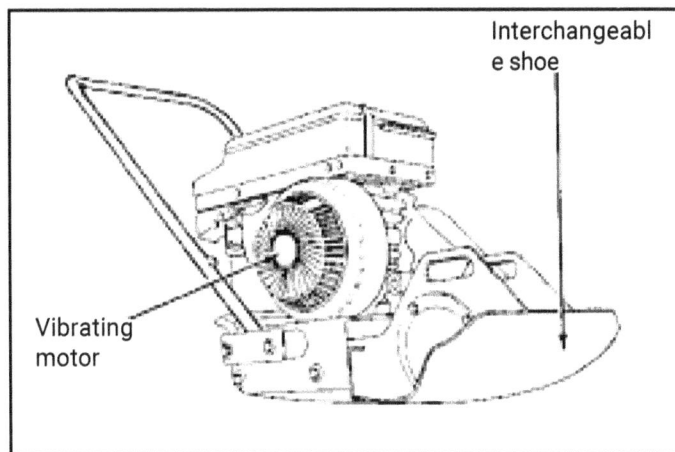

Figure 11-12. Jay Tamper

EQUIPMENT TESTING

11-22. Since the use of several types of compaction equipment overlaps, it is good to use a test strip to make the final determination of the most efficient compactor and compaction procedures. Locate a test strip adjacent to the project site. The test strip provides an evaluation/validation of the proposed construction procedures. Information obtained from atest strip includes the—

- Most effective type of compaction equipment.
- Optimum depth of lift.
- Optimum compactor speed.
- Number of passes required.
- Amount of ballast required.
- Vibration frequency required.

11-23. To make the initial selection of compaction equipment, place a 6-inch (uncompacted depth) lift of the material and run each piece of equipment over a specific length of the strip a predetermined number of passes (usually three). After compacting all the lengths, perform a density test to determine which roller gives the best results. After determining the most effective type of equipment, use additional test strips to determine the most appropriate lift thickness, the required number of passes, the optimum compactor speed and, if using a vibratory roller, the resonant frequency.

PRODUCTION ESTIMATES

11-24. Use the following formula to determine compactor production in CCY per hour.

$$\text{Production (CCY per hour)} = \frac{16.3 \times W \times S \times L \times E}{N}$$

where—

16.3 = constant for converting the factors in feet, mph, and inches to CCY

W = compacted width per pass, in feet

S = compactor speed, in mph

L = compacted lift thickness, in inches

E = efficiency

N = number of passes required

11-25. The accuracy of the compaction estimate depends on the project officer's skill in estimating the speed, the lift thickness, and the number of passes required to attain the required density. Normally, this information is determined from on-site test strips. The efficiency of a daytime compaction effort is typically 50 minutes per hour. Reduce the efficiency to 45 minutes per hour for nighttime compaction. Typical compactor speeds are given in *Table 11-4*. These speeds should be matched with the data in *Tables 11-2* and *11-3*, *pages 11-8* and *11-9*. When calculating production estimates, use the average speed of the compactor based on the individual speed for each pass.

Table 11-4. Typical Operating Speeds of Compaction Equipment

Compactor	Speed (mph)
Sheepsfoot, crawler, towed	3-5
Sheepsfoot, wheel-tractor, towed	5-10

Tamping foot:
    First two or three passes
    Walking out
	3-5
	8-10

Heavy pneumatic	3-5
Multitired pneumatic	5-15
Dual-drum vibratory	2-4
Smooth-drum vibratory	2-4

Vibratory:
    Plate
    Roller
	0.6-1.2
	1

EXAMPLE

Use of a test strip determined that it will take five passes to achieve the required density using a tamping-foot roller. The following speeds were achieved:

First pass = 4 mph
Second pass = 4 mph
Third pass = 5 mph
Fourth pass (walking out) = 8 mph
Fifth pass (walking out) = 9 mph

Determine the average speed.

$$\text{Average speed (mph)} = \frac{4+4+5+8+9}{5 \text{ passes}} = 6 \text{ mph}$$

EXAMPLE

What is the estimated production rate (CCY per hour) for a tamping-foot roller with a compaction width of 5 feet? The following information was obtained from a test strip at the project:

Compacted lift thickness = 6 inches
Average speed = 6 mph
Number of passes = 5
Efficiency factor = 0.83

$$\text{Production (CCY per hour)} = \frac{16.3 \times 5 \times 6 \times 6 \times 0.83}{5 \text{ passes}} = 487 \text{ CCY per hour}$$

11-26. Determine the total number of compactors required on the project, using the following formula.

$$\text{Compactors required} = \frac{\text{amount of fill delivered (LCY per hour)} \quad \text{soil conversion factor (LCY:BCY)}}{\text{compactor production (CCY per hour)}}$$

NOTE: Refer to *Table 1-1, page 1-4,* for soil conversion factors.

EXAMPLE

How many compactors are required on the project (previous example) if 1,500 LCY of blasted rock is delivered per hour?

$$\text{Compactors required} = \frac{1,500 \text{ LCY per hour} \times 0.87 \text{ (soil conversion factor)}}{487 \text{ CCY per hour}} = 2.7 \text{ compactors, round up to 3}$$

HORIZONTAL CONSTRUCTION PHASES

11-27. To satisfy the project's density requirement, it is essential to understand the various phases of a horizontal construction project. These include preparing the subgrade; placing and spreading fill material; compacting fill material for the subgrade and base courses; and performing finishing and surfacing operations (discussed in *Chapter 12*).

Preparing Subgrade

11-28. The term subgrade describes the in-place soil on which a road, an airfield, or a heliport is built. Subgrade includes the soil to the depth that may affect the structural design of the project or the depth at which the climate affects the soil. Depths up to 10 feet may be considered subgrade for pavements carrying heavy loads. The quality and natural density of this material dictate what action(s) to take to prepare the subgrade. For example, a highly organic subgrade material may have to be totally removed and replaced with a higher quality, select material. In most situations this is not the case. Often the in-place material is suitable, but requires some degree of compactive effort to achieve the required density.

11-29. Heavy, pneumatic-tired rollers are preferred for subgrade compaction because of their capability to compact the soil to depths up to 18 inches. Often it is necessary to scarify the top 6 inches of the material to adjust the moisture content. Then the material can be recompacted to a higher density than could be achieved at the soil's natural moisture content. This process is known as scarify and compact in place (SCIP). The tamping-foot roller is good for this operation since it loosens the material, yet compacts it as it walks out of the material. The tamping-foot roller also helps to break up oversize material or rock in the area. If using a tamping-foot roller to prepare the subgrade, also use a pneumatic-tired or smooth-drum roller to compact the top 1 to 2 inches of the final material lift or to seal the lift surface if expecting rain.

11-30. Shaping and sealing the surface protects the subgrade from the damaging effects of water infiltration. If a lift has been sealed, scarify the top 1 or 2 inches before placement of a succeeding lift. This ensures a good bond between the lifts. If the subgrade material is sand, a vibratory compactor would provide the most effective compactive effort.

Placing and Spreading Fill Material

11-31. After preparing the subgrade, bring in fill material to form the subbase and base courses for the project. When placing fill, it is important to spread the material in uniform layers and to maintain a reasonably even surface. The thickness of the layers is dependent on the desired compacted lift thickness. The thickness of the uncompacted lift is normally 1 1/2 to 2 times the final compacted lift. For example, place fill in 9- to 12-inch lifts to achieve a compacted lift thickness of 6 inches. Place the fill material with a scraper or a dump truck and spread it with a dozer or grader. When spreading material on a prepared subgrade, spread the material from the farthest point from the source to the nearest or vice versa. The advantages of spreading fill from the farthest point to the nearest are as follows:

- The hauling equipment will further compact the subgrade.
- Previously undetected weaknesses in the subgrade will become apparent.
- Hauling will not hinder spreading or compacting operations.

11-32. On the other hand, spreading from the nearest point to the farthest point has the advantage of haul equipment traveling over the newly spread material. This compacts the material and greatly reduces the overall compactive effort required.

Compacting Fill Material

11-33. Before beginning compaction operations, the project officer must determine the moisture content of the fill and compare it to the acceptable moisture range for that material. If the moisture content is below the acceptable range, add water to the fill. If the moisture content is higher than the acceptable range, use one of the previously discussed methods to dry the soil. After achieving the appropriate moisture content, begin compaction operations. The base course functions as the primary load-bearing component of the road, ultimately providing the pavement (or other surface) strength. Base-course material is, therefore, higher quality material than either the subgrade or the subbase fill. Base course normally consists of well-graded granular materials that have a liquid limit less than 25 percent and a plastic limit less than 6 percent. The thickness of the base course is dependent on the strength of the subgrade. Smooth-drum vibratory rollers are ideal for base-course compaction. A dual-drum roller can also be used for base-course compaction.

OPERATING HINTS

Compacting Against Structures

11-34. Jay tampers and pneumatic backfill tampers are specifically designed for use in confined areas such as against existing structures. If these tampers are not available, use a roller to achieve satisfactory results. If space permits, run the roller parallel to the structure. If it is necessary to work perpendicular to the structure, place fill material (sloped to the height of the roller axle) against the structure. Apply compactive effort against this excess fill material. Take care to avoid damaging the existing structure.

Aerating Materials

11-35. When using the roller to aerate soils, travel at the highest practical speed. High speeds tend to kick up the material, which is the objective in this case instead of density.

Overlapping Passes

11-36. To eliminate noncompacted strips, each pass with the roller should overlap the preceding pass by at least 1 foot.

Turning

11-37. Make gradual turns at the end of each pass. This prevents surface damage to the lift and eliminates the possibility of damaging a towed roller's tongue with the tracks of the towing tractor.

Chapter 12

Road Surfacing

This chapter covers surface treatments and the road-surfacing equipment used to apply these treatments. Surface treatments include the following: seal coat, sprayed bituminous material with cover aggregate (single and multiple surface treatments), and asphalt emulsions or slurry seal. Paving and surfacing operations are discussed in FM 5-436.

SURFACE TREATMENT

12-1. The term *surface treatment* covers a wide range of applications and materials, with or without aggregate, applied to the top of a road or pavement. Surface treatments with aggregates can be—

- **Single surface.** A single surface treatment is a sprayed bituminous material with an aggregate cover that is one stone in depth. Use a single surface treatment as a wearing and waterproofing treatment. This type of surfacing requires a minimum of equipment, materials, and time.
- **Multiple surface.** Repeating the single surface-treatment process results in what is referred to as a multiple surface treatment. A multiple surface treatment provides a denser wearing and waterproofing course and the multiple layers add strength to the road structure.

An initial single or double surface treatment may, when time and material are available, be supplemented by additional surface treatments or thicker pavement structures. This process is referred to as *stage construction*.

SURFACING EQUIPMENT

SWEEPER

12-2. A sweeper is a tractor-type machine with one or more brooms for removing dust from the surface of the existing roadway before laying new asphalt. Remove dust or dirt to ensure proper bonding between the new asphalt and the base course or old pavement. When surfacing a prepared base course, remove the dust layer either by sweeping with the sweeper or by wetting the base course and recompacting. Before using a sweeper, always inspect for excess bristle wear and check that the power drive on all brooms is operating properly.

ASPHALT DISTRIBUTOR

12-3. The Army asphalt distributor *(Figure 12-1)* is a truck-mounted unit. It has a 1,500-gallon insulated storage tank with a low-pressure heating system, a hydraulic-powered pumping unit mounted on the rear, and an adjustable spray bar for distributing bituminous material. To produce a uniform application, an asphalt distributor requires constant attention. It is critical that the asphalt heater and pump be well maintained. Calibrate all gauges and measuring devices properly (pump tachometer, measuring stick, thermometers, bitumeter, and so on). Use clean spray bars and nozzles and set them at the proper height above the surface receiving the application.

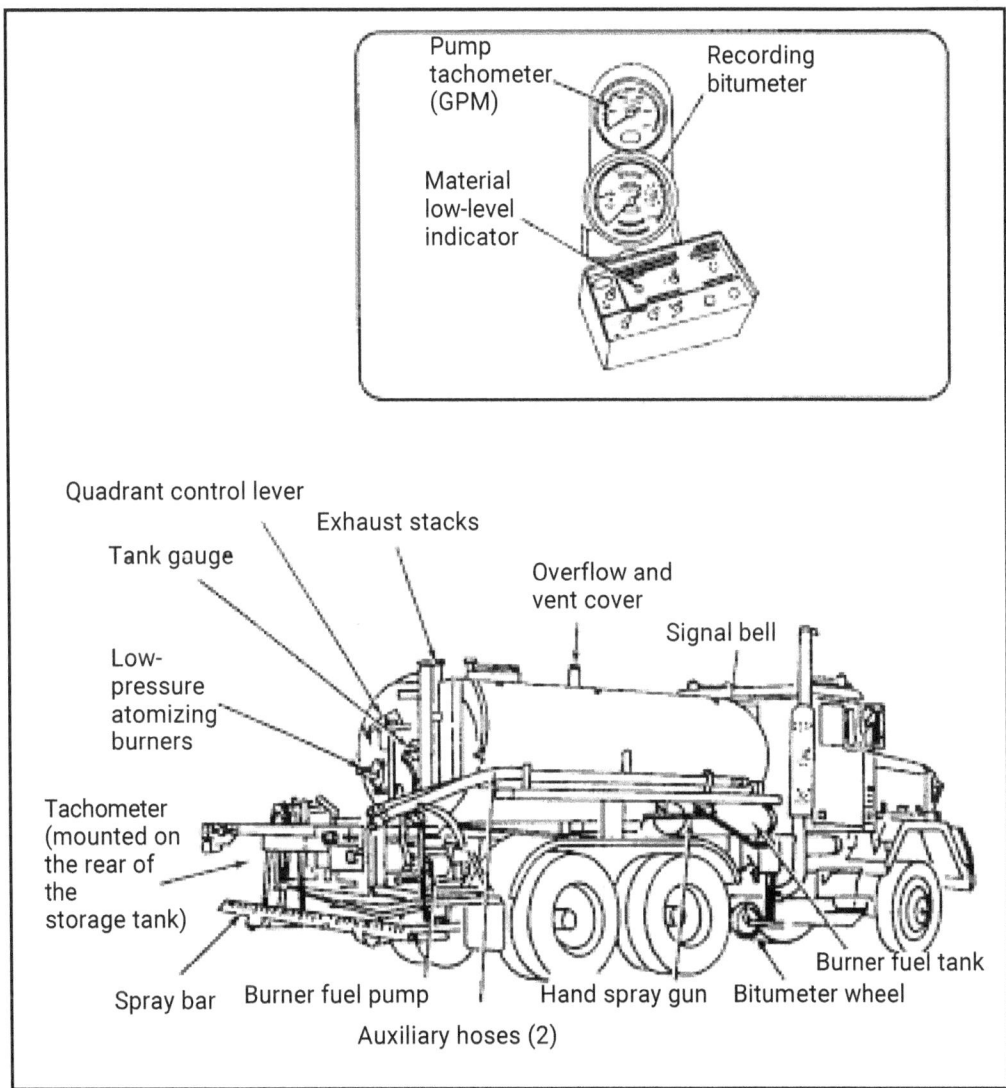

Figure 12-1. Army Asphalt Distributor

Heating System

12-4. The system's two diesel burners each have an air blower that provides low-pressure air for atomizing burner fuel. The burners emit flame into two U-shaped fire tubes located in the bottom of the asphalt storage tank. Maintain 3 inches of asphalt cover over these fire tubes to prevent overheating and possible fire or explosion. Perform heating only in a well-ventilated area with the distributor truck level and at a complete stop. *Table 12-1, page 12-4,* is a guide for asphalt spraying temperatures.

> **WARNING**
>
> **Exercise extreme caution while heating the asphalt to prevent damage to the heating system and possible fires and explosions.**

Spraying System

12-5. The spraying system consists of piping, an adjustable spray bar, a bitumen-pump tachometer, and a bitumeter. Use *Table 12-2, page 12-5,* to determine the proper settings for a desired application rate. The application rate is controlled by —

- The length of the spray bar.
- The bitumen pump output. The pump tachometer dial in the cab of the truck (*Figure 12-1*) registers the pump output in GPM from the tachometer mounted on the rear of the storage tank.
- The forward speed of the distributor truck. The bitumeter mounted inside the truck cab monitors the forward speed in fpm.

12-6. **Spray Bar.** The spray bar may be full-circulating or noncirculating, depending on the model of the distributor. The bar can be adjusted to provide coverage from 8 to 24 feet in width. Along the bar there are a series of nozzles with hand-operated valves to control the flow of asphalt. Although the spray bar may have either 1/8- or 3/16-inch nozzles, the 1/8-inch nozzle is appropriate for most applications. For a uniform application, make sure that the spray bar is at the proper height and that all nozzles are the same size and free of obstructions.

- **Spray-bar height.** To achieve a uniform double or triple overlap for a single-layer application, the height of the spray bar must be properly adjusted above the pavement. It is important to maintain the correct height during the entire application. If the spray bar is too low or too high, streaking will result.
- **Nozzle spacing.** Spray bars are usually constructed with either a 4- or 6-inch nozzle spacing. With 4-inch nozzle spacing, the best application results are attained using a triple lap of the spray fans (*Figure 12-2, page 12-6*). To determine the proper spray-bar height for triple-lap coverage, open every third nozzle on the spray bar. Raise the bar until there is single-layer coverage along the entire length of the bar. Do a visual inspection to determine this. When all the nozzles are opened, this height will furnish a triple lap of the spray fans. Use the same procedure for double-lap coverage except only open every other nozzle. When using 6-inch nozzle spacing, the height of the bar necessary to give a triple-lap coverage will frequently allow wind distortion of the spray fans, resulting in a nonuniform application. Therefore, with 6-inch nozzle spacing, only a double lap may be achieved.

Table 12-1. Typical Pug Mill and Spraying Temperatures for Asphalts (Degrees Fahrenheit

Type and Grade of Asphalt		Pug Mill Mixture Temperatures [1]		Spraying Temperatures [5]		Notes
		Dense-Graded Mixes	Open-Graded Mixes	Road Mixes	Surface Treatments	
Asphalt cements	AC-2.5	235-280	180-250	æ	270+	Temperatures for asphalt cements and cutback asphalts are guides only.
	AC-5	250-295	180-250	æ	280+	
	AC-10	250-315	180-250	æ	280+	
	AC-20	265-330	180-250	æ	295+	
	AC-40	270-340	180-250	æ	300+	[1]Temperature of mixture immediately after discharge from the pug mill rather than temperature of asphalt cement or cutback asphalt.
	AR-1,000	225-275	180-250	—	275+	
	AR-2,000	275-325	180-250	—	285+	
	AR-4,000	275-325	180-250	—	290+	
	AR-8,000	275-325	180-250	—	295+	
	AR-16,000	300-350	180-250	—	—	
	200-300 pen	325-305	180-250	—	265+	[2]Application temperatures may, in some cases, be above the flash point of the material. Exercise caution to prevent fire or explosion.
	120-150 pen	245-310	180-250	—	270+	
	85-100 pen	250-325	180-250	—	280+	
	60-70 pen	265-335	180-250	—	295+	
	40-50 pen	270-350	180-250	—	300+	
Cutback asphalts (RC, MC, SC) [2]	30 (MC only)	—	—	—	85+	
	70	—	65+	120+	—	
	250	135-175 [3]	—	105+	165+	
	800	165-210 [3]	—	135+	200+	[3]Rapid curing grades are not recommended for hot pug mill mixtures.
	3,000	180-240 [3]	—	—	230+	
Emulsified asphalts	RS-1	—		—	70-140	
	RS-2	—		—	125-185	
	MS-1	50-160 [4]		70-160	—	[4]Temperature of the emulsified asphalt in the pug mill mixture.
	MS-2	50-160 [4]		70-160	—	
	MS-2h	50-160 [4]		70-160	—	
	SS-1	50-160 [4]		70-160	—	[5]The maximum temperature (asphalt cement and cutback asphalt) must be below the temperature at which fogging occurs.
	SS-1h	50-160 [4]		70-160	—	
	CRS-1	—		—	125-185	
	CRS-2	—		—	125-185	
	CMS-2	50-160 [4]		70-160	—	
	CMS-2h	50-160 [4]		70-160	—	
	CSS-1	50-160 [4]		70-160	—	
	CSS-1h	50-160 [4]		70-160		

Legend :

AC = viscosity-graded asphalt cement
AR = viscosity-graded, aged-residue asphalt cement
CMS = cationic-medium cement
CRS = cationic rapid-setting
CSS = cationic slow-setting

MC = medium curing
MS = medium setting
pen = penetration-graded asphalt
RC = rapid curing
RS = rapid setting
SS = slow setting

)

Table 12-2. Asphalt-Distributor Application Rates

Gallons Per Sq Yd	Nozzle Size (Inch)	Bitu-meter Reading	Pump Tachometer Readings (GPM) — Spray-Bar Length (Feet)															
			8	9	10	11	12	13	14	15	16	17	18	19	20	21	22	24
0.10	1/8	900	80	90	100	110	120	130	140	150	160	170	180	190	200	210	220	240
	3/16	1350	120	135	150	165	180	195	210	225	240	255	270	285	300			
0.20	1/8	450	80	90	100	110	120	130	140	150	160	170	180	190	200	210	220	240
	3/16	675	120	135	150	165	180	195	210	225	240	255	270	285	300			
0.25	1/8	360	80	90	100	110	120	130	140	150	160	170	180	190	200	210	220	240
	3/16	540	120	135	150	165	180	195	210	225	240	255	270	285	300			
0.30	1/8	300	80	90	100	110	120	130	140	150	160	170	180	190	200	210	220	240
	3/16	450	120	135	150	165	180	195	210	225	240	255	270	285	300			
0.40	1/8	225	80	90	100	110	120	130	140	150	160	170	180	190	200	210	220	240
	3/16	340	120	135	150	165	180	195	210	225	240	255	270	285	300			
0.50	1/8	180	80	90	100	110	120	130	140	150	160	170	180	190	200	210	220	240
	3/16	270	120	135	150	165	180	195	210	225	240	255	270	285	300			
0.60	1/8	150	80	90	100	110	120	130	140	150	160	170	180	190	200	210	220	240
	3/16	225	120	135	150	165	180	195	210	225	240	255	270	285	300			
0.70	1/8	130	80	90	100	110	120	130	140	150	160	170	180	190	200	210	220	240
	3/16	195	120	135	150	165	180	195	210	225	240	255	270	285	300			
0.75	1/8	120	80	90	100	110	120	130	140	150	160	170	180	190	200	210	220	240
	3/16	180	120	135	150	165	180	195	210	225	240	255	270	285	300			
0.80	1/8	110	80	90	100	110	120	130	140	150	160	170	180	190	200	210	220	240
	3/16	170	120	135	150	165	180	195	210	225	240	255	270	285	300			
0.90	1/8	100	80	90	100	110	120	130	140	150	160	170	180	190	200	210	220	240
	3/16	150	120	135	150	165	180	195	210	225	240	255	270	285	300			
1.00	1/8	90	80	90	100	110	120	130	140	150	160	170	180	190	200	210	220	240
	3/16	135	120	135	150	165	180	195	210	225	240	255	270	285	300			
1.10	1/8	80	80	90	100	110	120	130	140	150	160	170	180	190	200	210	220	240
	3/16	120	120	135	150	165	180	195	210	225	240	255	270	285	300			
1.20	1/8	75	80	90	100	110	120	130	140	150	160	170	180	190	200	210	220	240
	3/16	110	120	135	150	165	180	195	210	225	240	255	270	285	300			
1.25	1/8	70	80	90	100	110	120	130	140	150	160	170	180	190	200	210	220	240
	3/16	105	120	135	150	165	180	195	210	225	240	255	270	285	300			
1.50	1/8	60	80	90	100	110	120	130	140	150	160	170	180	190	200	210	220	240
	3/16	90	120	135	150	165	180	195	210	225	240	255	270	285	300			
1.75	1/8	50	80	90	100	110	120	130	140	150	160	170	180	190	200	210	220	240
	3/16	80	120	135	150	165	180	195	210	225	240	255	270	285	300			
2.00	1/8	45	80	90	100	110	120	130	140	150	160	170	180	190	200	210	220	240
	3/16	70	120	135	150	165	180	195	210	225	240	255	270	285	300			
2.50	1/8	35	80	90	100	110	120	130	140	150	160	170	180	190	200	210	220	240
	3/16	55	120	135	150	165	180	195	210	225	240	255	270	285	300			
3.00	1/8	30	80	90	100	110	120	130	140	150	160	170	180	190	200	210	220	240
	3/16	45	120	135	150	165	180	195	210	225	240	255	270	285	300			

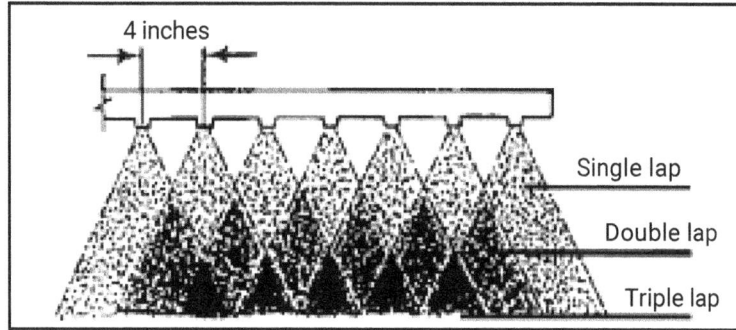

Figure 12-2. Perfect Triple Lap

- **Nozzle angle setting.** To attain a good edge, the end nozzles are often set at a different angle (60 ° with respect to the spray bar) from the other nozzles. Do not permit this practice because it produces a fat streak on the edge and robs the adjacent spray fan of the lap from this nozzle. A curtain on the end of the bar or a special end nozzle with all nozzles set at the same angle gives more uniform coverage and makes a better edge *(Figure 12-3)*.

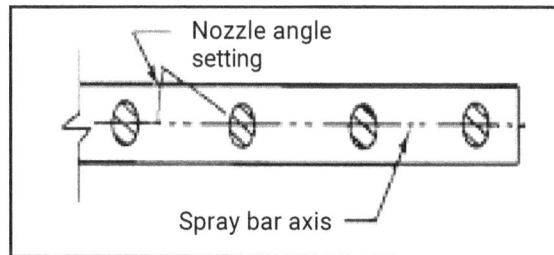

Figure 12-3. Proper Nozzle Angle Setting

12-7. **Bitumeter.** The bitumeter *(Figure 12-1, page 12-2)* has a small rubber-tired wheel mounted on a retractable frame under the distributor. A cable leads from the wheel to a dial in the truck cab. The dial registers the distributor's rate of travel in fpm and the travel distance in feet. Some bitumeters, besides providing travel distance, correspondingly register the application rate in gallons per square yard. Check the bitumeter at regular intervals to ensure accurate registering of speeds. The bitumeter check procedure is as follows:

Step 1. Mark off a distance of 500 to 1,000 feet on a straight, level surface.

Step 2. Drive the distributor at a constant speed over this length while timing the trip with a stopwatch.

Step 3. Determine the distributor speed in fpm and compare this calculation with the bitumeter reading recorded during the run.

Step 4. Repeat this procedure for a number of different speeds, bracketing the speed to be used for spraying.

Step 5. Tabulate or plot on a graph errors found at various speeds so a correction can be readily applied when using the distributor.

NOTE: Keep the bitumeter wheel clean to ensure accurate registering of the truck's speed. Asphalt buildup on the wheel causes errors.

AGGREGATE SPREADER

12-8. Use the towed aggregate spreader*(Figure 12-4)* to apply aggregate to freshly sprayed asphalt surfaces. It has a 2.5-cubic-yard capacity charging hopper, an aggregate feed roller, an adjustable strike-off gate, traction tires, a towing tongue, and a travel axle. During spreading operations the spreader connects to the dump truck with a two-point, quick-coupling hitch but is supported by its traction tires. It can apply aggregate when being towed or when pushed by the dump truck that is supplying aggregate to the charging hopper. Push the spreader with the dump truck for surface-treatment operations. This allows the aggregate spreader and the dump truck to ride on the freshly laid aggregate. As the aggregate spreader moves along the project, the rotation of the traction tires drives the aggregate feed roller. Control the application rate (pounds per square yard) of the aggregate with the adjustable gate setting. The application width ranges from 4 to 8 feet, in 1-foot increments. Control the width of application by installing blocking plates inside the charging hopper. The blocking plates merely block off a segment of the gate opening.

NOTE: It is important to spread the aggregate immediately after application of the asphalt, before the asphalt cools.

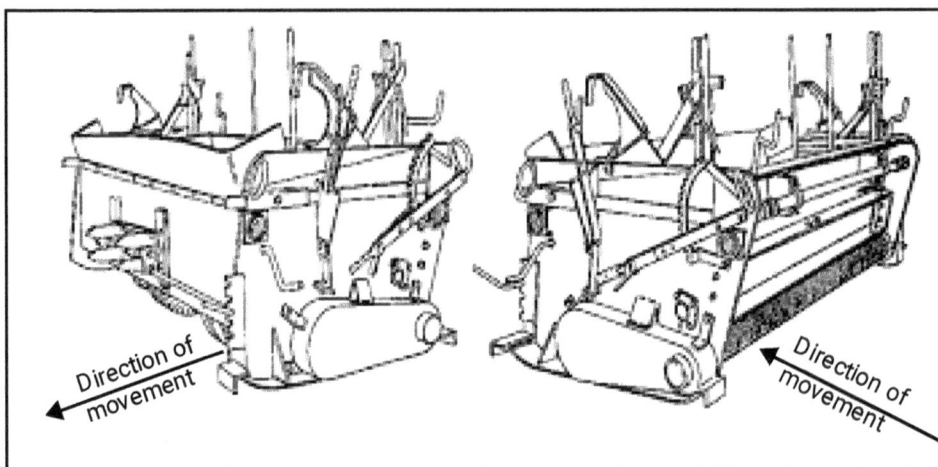

Figure 12-4. Towed Aggregate Spreader

ROLLER

12-9. All rollers must be in good mechanical condition and able to start, stop, and reverse smoothly. They must have a water-sprinkling system to keep the drums or tires wet so that asphalt will not stick to the drum surface and scar the new surface. Operate the rollers on the asphalt with the drive wheel toward the spreader or asphalt paver. This pulls the mix under the roller and imparts a smooth surface. Placing the rollers on the asphalt with the guide roller toward the spreader will push the hot asphalt forward *(Figure 12-5, page 12-8)*. This causes the material to mound up in front of the roller. Eventually, the roller passes over the mound, leaving a bump in the finished pavement. In most cases, subsequent rolling cannot correct this bump.

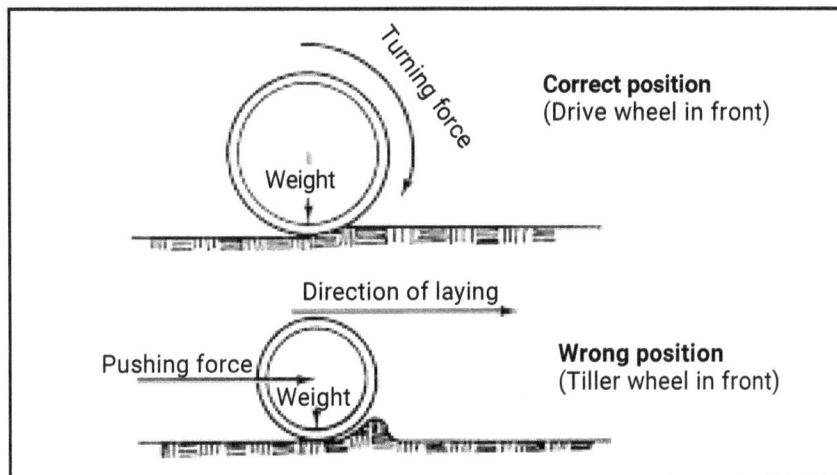

Figure 12-5. Correct Roller Drive-Wheel Position

Dual-Drum Vibratory Roller

12-10. A 7- to 14-ton, dual-drum vibratory roller can be used for compacting surface treatments, normally after application of the second layer of aggregate. However, the roller should not be so heavy or the vibration amplitude set so high that it crushes the aggregate particles.

Pneumatic-Tired Roller

12-11. A pneumatic-tired roller is the preferred equipment for compacting bituminous surface treatments. See *Chapter 11* for information on pneumatic-tired rollers.

IN-PLACE MIXING EQUIPMENT

12-12. The Army has two pieces of equipment for in-place mixing of asphalt or other stabilizing agents—grader and stabilizer-mixer.

GRADER

12-13. Use a grader to spread the aggregate, either from piles or by scarifying in-place material. After applying the liquid asphalt with the distributor, use the grader to mix the asphalt and aggregate thoroughly. For mixing, pitch the blade all the way forward and angle it to side cast the windrow. Normally, it takes five or six passes with the grader to thoroughly mix the materials after each asphalt application. See *Chapter 4* for additional information on asphalt mixing with a grader.

STABILIZER MIXER

12-14. The stabilizer mixer is preferred over a grader for road mixing materials. It is faster and provides better control of the asphalt/aggregate ratio, producing a thoroughly mixed material. First, use the mixer to loosen in-place materials and to blend select aggregate. The asphalt distributor supplies asphalt directly to the mixer pump through a flexible metal hose. When using this hose, connect the two units with a safety chain to prevent damage to the hose. Spray the asphalt onto the aggregate with a spray bar mounted under the mixer hood. The spraying takes place while the mixer is moving and the tines are rotating. This procedure completely blends the asphalt and aggregate. See *Chapter 11* for additional information on stabilizer mixers.

BITUMEN HANDLING AND DEDRUMMING EQUIPMENT

ASPHALT MELTER

12-15. The asphalt melter *(Figure 12-6)* melts asphalt from 55-gallon drums and heats it to a prescribed pumping temperature. The upper section of the melter consists of dual, dedrumming tunnels; the lower section consists of an asphalt melting and storage tank. In dedrumming operations, cut away one end of each 55-gallon drum. Then place the drum, open end down, in the dedrumming tunnel. Use an oil heater in conjunction with the asphalt melter. Hot-oil coils, running along the sides and bottom of the tunnel, provide the heat to melt the asphalt in the drums. Additional hot-oil coils, running throughout the storage tank, bring the asphalt to mixing temperature. *Table 12-1, page 12-4,* is a guide for asphalt mixing temperatures in pug mills.

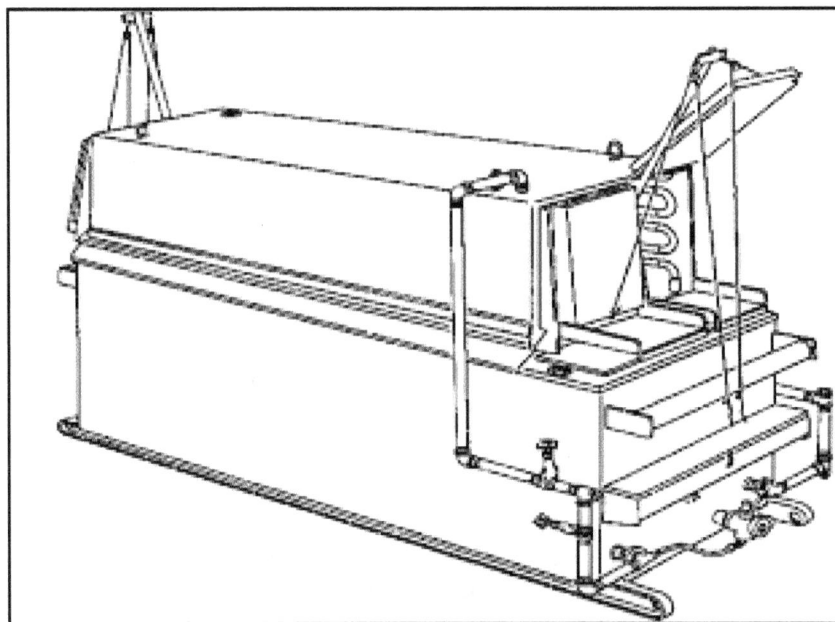

Figure 12-6. Asphalt Melter

OIL HEATER

12-16. The oil heater heats and then circulates the hot oil to systems such as an asphalt melter or a storage tank. A diesel burner provides the heat to raise the temperature of the circulating oil. A pump circulates the hot oil to the systems. Set the heating requirements on the heater's temperature controller. From that point on, a programmed circuit automatically controls the operation with a check-and-balance safety procedure. If the system malfunctions, safety devices cause the unit to shut down. The oil heaters come mounted on liquid-asphalt storage-tank trailers.

ASPHALT KETTLE

12-17. The 165-gallon asphalt kettle *(Figure 12-7)* is designed primarily for surface patching and maintenance operations. The trailer-mounted tank consists of an outer shell encompassing a 165-gallon capacity storage and heating tank. A removable distillate burner, mounted inside the outer shell, provides heat to the kettle. A flue carries off the exhaust gases. A small, two-cylinder gasoline engine provides power to the asphalt pump. Asphalt is applied to small patching projects through a flexible hose and a handheld spray assembly.

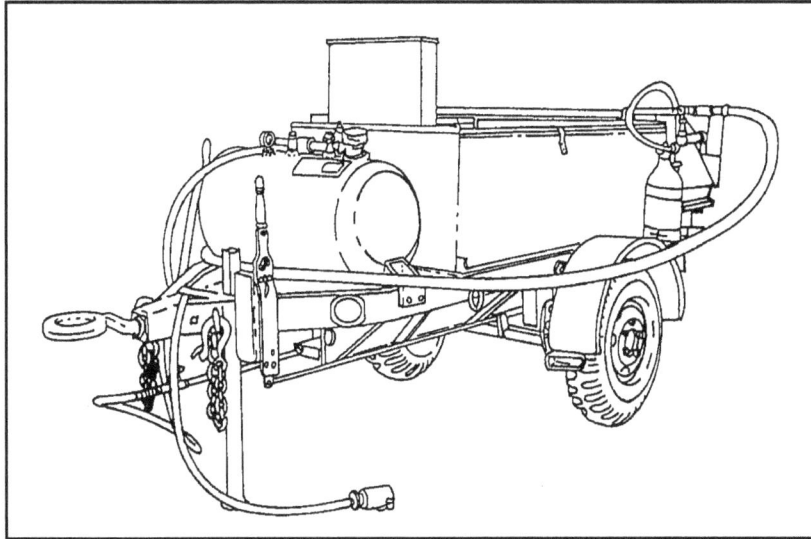

Figure 12-7. Asphalt Kettle

SUPPORT EQUIPMENT

12-18. Use a rough-terrain forklift to transport and lift asphalt drums and mineral filler. See *Chapter 6* for information on forklifts. Use a loader to maintain aggregate stockpiles and for other material-handling operations. See *Chapter 5* for information on loaders.

Chapter 13

Safety

Supervisors are responsible for ensuring that personnel follow safety standards. Time is usually the controlling factor in construction operations in the theater of operations. The necessity for economy of time, coupled with the temporary nature of much of the work, sometimes results in safety precautions that are substantially lower than those used in civilian practice, but this does not mean safety can be ignored. Construction equipment, except for dump trucks, is designed to accommodate only the operator.

SAFETY PROGRAM

13-1. Do not construe the lack of documentation of hazards as an indication of their nonexistence or insignificance. Where safety precautions are necessary but are not documented, or where existing precautions are judged to be inadequate, the commanding officer must issue new or supplementary warnings. Each job has its own particular safety hazards. Identify dangers and prepare a safety program to reduce or eliminate all hazards. Supervisors must conduct all operations following the guidance in the safety program. For help in making safety decisions, review the safety information in United States (US) Army Corps of Engineers Manual (EM) 385-1-1. Additionally, operators' manuals give do's and don'ts, cautions, and safety warnings. Check them!

GENERAL SAFETY RULES

13-2. The appropriate chapters of this manual identify safety rules for specific equipment. Also, check applicable technical and operator manuals prior to operating all equipment. Some general safety rules are as follows:

- Inspect equipment before use, and periodically on a regular basis.
- Ensure that mechanized equipment is operated by qualified and authorized personnel only.
- Use seat belts when they are available.
- Provide barriers to prevent personnel from walking under loading equipment that has a hoist or lift capability.
- Operate equipment in a manner that will not endanger persons or property.
- Observe safe operating speeds.
- Shut down and turn off the engine when equipment is unattended.
- Stop the equipment completely (apply the parking brake if available) before mounting or dismounting.

- Do not operate any machinery or equipment for more than 10 consecutive hours without an 8-hour rest interval.
- Post the safe load capacities at the operator's position on all equipment not rigged to prevent overloading.
- Post the safe operating speeds at the operator's position on all equipment not having a speed governor.
- Mount a ROPS on equipment whenever it is available. Do not make any modifications to the ROPS.
- Ensure that only the operator is on the equipment while it is running. Supervisors can authorize exceptions in emergency situations, some training situations, and when required for maintenance.
- Shut down and turn off the engine when refueling motor vehicles and mechanized equipment.

OPERATOR INDOCTRINATION

13-3. Teach operator safety practices during machine operation training. Indoctrinate operators about job hazards and ways to reduce or avoid them. Instruct operators to observe fellow workers and to warn them when they get into dangerous situations. Personnel should receive continuing instruction during the project to ensure that they appreciate the job hazards and understand the objectives of the safety program. Horseplay and practical jokes create very dangerous situations on the work site.

OPERATOR QUALIFICATIONS AND REQUIREMENTS

13-4. Construction and material-handling equipment operators must be tested and licensed according to Army Regulation (AR) 600-55. An apprentice or applicant can only operate equipment under the direct supervision of a licensed operator. Do not permit anyone who is not physically or mentally capable to operate any equipment. Operators are responsible for the safe operation of their equipment and for the safety of their passengers and cargo.

EQUIPMENT INSPECTION

13-5. Before using a machine, a qualified, licensed operator should inspect and test the equipment to determine its safe operating condition. Equipment-operator maintenance checks, service charts, and common sense ensure safe operation and proper maintenance. Tag any unsafe machinery or equipment "Out Of Service, Do Not Use" at the operator's position, to prevent its use until repaired. Ensure that the equipment's safety features (backup alarms, lights, and so on) are operational.

REPAIRS AND MAINTENANCE

13-6. Besides the following items, check the unit's SOP for special repair and maintenance procedures.

- Shut down or lock out equipment controls while a machine is being repaired, adjusted, or serviced.
- Position the equipment in a place, away from the project area, that is safe for the mechanic to work.

- Crib or block suspended machinery, equipment, or parts, and machines held apart by slings, hoists, or jacks. Do not work underneath or between items not properly blocked.
- Lower blades, bowls, hooks, buckets, and forks to the ground or onto suitable blocking material when equipment is undergoing maintenance or repairs.

GUARDS AND SAFETY DEVICES

13-7. Equipment guards, appliances, and similar devices are for the protection of personnel. Do not remove these devices or render them ineffective. During repairs, lubrication, or adjustments, remove guards temporarily, but only after shutting off the power. Replace all guards and devices immediately after completing repairs and adjustments. All equipment should have properly working warning devices, such as backup alarms and turn signals.

SIGNALS

13-8. Provide a warning device or a signal person wherever there is possible danger to fellow workers or others from moving equipment. If the tactical situation allows, the signal person should wear a reflective, orange vest.

ROPES, CABLES, AND CHAINS

13-9. *Table 13-1* and *Tables 13-2* and *13-3, page 13-4,* give load limitations and safety factors for ropes, cables, and chains. Ropes, cables, and chains used in construction operations present some of the largest potential safety hazards. To eliminate these hazards—

- Perform periodic inspections of their physical condition.
- Use only according to manufacturers' recommended procedures or within the safe limits recommended by the manufacturers of the equipment.
- Wear leather gloves when handling wire rope.

Table 13-1. Breaking Strengths of 6 x 19 Standard Wire Hoisting Rope

Diameter (Inches)	Approximate Weight Per 100 Feet (Pounds)	Breaking Strength in Tons (2,000-Pound Ton)		
		Mild Plow Steel	Plow Steel	Improved Plow Steel
1/4	10	2.07	2.39	2.74
3/8	23	5.00	5.50	6.30
1/2	40	8.50	9.40	10.80
5/8	63	13.10	14.40	16.60
3/4	90	18.70	20.60	23.70
7/8	123	25.40	28.00	32.20
1	160	33.00	36.50	42.00
1 1/8	203	41.50	46.00	53.00
1 1/4	250	57.00	56.50	65.00
1 1/2	360	72.50	80.50	92.50

NOTES:
1. A 6 x 19 rope is composed of 6 strands of 19 wires each. The strength of wire rope varies slightly with the strand construction and the number of strands.
2. The maximum allowable load is the breaking strength divided by the appropriate safety factor. The safety factors are shown in *Table 13-3, page 13-4.*

Table 13-2. Breaking Strengths and Safe Loads for Manila and Sisal Rope

Nominal Diameter (Inches)	Circumference (Inches)	Weight Per 100 Feet (Pounds)	Number 1 Manila		Sisal	
			Breaking Strength (Pounds)	Safe Load (Pounds) (Safety Factor = 4)	Breaking Strength (Pounds)	Safe Load (Pounds) (Safety Factor = 4)
1/4	3/4	2.00	540	140	440	120
3/8	1 1/8	4.10	1,260	320	1,020	260
1/2	1 1/2	7.50	2,640	660	2,120	530
5/8	2	13.30	4,400	1,100	3,520	880
3/4	2 1/4	16.70	5,400	1,350	4,320	1,080
7/8	2 3/4	18.60	7,700	1,920	6,160	1,540
1	3	27.00	9,000	2,250	7,200	1,800
1 1/8	3 1/2	36.00	12,000	3,000	9,600	2,400
1 1/4	3 3/4	41.80	13,440	3,360	10,800	2,700
1 1/2	4 1/2	60.00	18,500	4,620	14,800	3,700
1 3/4	5 1/2	89.50	26,500	6,620	21,200	5,300
2	6	108.00	31,000	7,750	24,800	6,200
2 1/2	7 1/2	135.00	46,500	11,620	37,200	9,300
3	9	242.00	64,000	16,000	51,200	12,800

NOTE: The above figures are for new rope used under favorable conditions. As rope ages or deteriorates, progressively reduce safe load values to one-half the values given. See safety factors shown in *Table 13-3*.

Table 13-3. Safety Factors for Ropes, Cables, and Chains

Type of Service	Wire Rope (Cable)	Manila or Sisal Rope	Chain
Guy lines	3.5	3.5	3.5
Miscellaneous hoisting equipment	5.0	7.0	5.0
Haulage lines	6.0	8.0	6.0
Overhead and gantry cranes	6.0	—	—
Jib and pillar cranes	6.0	—	—
Derricks	6.0	—	—
Small, electric or air hoists	7.0	—	7.0
Slings	8.0	10.0	8.0

NOTE: To determine the safe load for a single line, divide the breaking strength of the line by the applicable safety factor and multiply the result by an assumed efficiency factor of 80 percent for fittings.

INSPECTION

13-10. Inspect wire rope or cable at the time of installation and as part of daily operator maintenance. Remove the wire rope or cable from hoisting service when it is kinked, or when it has a number of broken wires in a strand. Remove the wire rope or cable when—

- Three broken wires are found in one strand of 6×7 wire rope.
- Six broken wires are found in one strand of 6×19 wire rope.
- Eight broken wires are found in one strand of 8×19 wire rope.
- Nine broken wires are found in one strand of 6×37 wire rope.

NOTE: Plainly mark defective wire rope as unsuitable for further use and dispose of properly.

SLINGS

13-11. Only qualified rigging personnel should attach slings to a load. When using slings, their fasteners, and other rigging attachments —

- Inspect daily for evidence of overloading, excessive wear, or damage. Replace defective slings or accessories.
- Make all eye splices *(Figure 13-1)* using the proper size of wire-rope thimbles.
- Place the U-bolt of the U-bolt clips on the dead (short) end of the wire rope (see *Figure 13-1*). Tighten the U-bolts immediately after initial load application and at frequent intervals thereafter. *Table 13-4* shows the number and spacing of clips and the proper torque to apply to the nuts of the clips (do not over torque). The minimum number of clips is three.

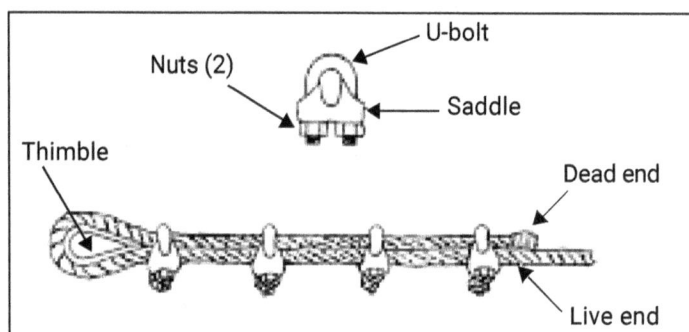

Figure 13-1. Eye splice

Table 13-4. Number and Spacing of U-Bolt clips for Wire-Rope Eye Connections

Wire-Rope Diameter		Nominal Clip Size (Inches)	Number of Clips	Spacing of Clips		Torque to be Applied to Nuts of Clips	
Inches	Millimeters			Inches	Millimeters	Foot-Pounds	(M-Kg) (0.1383)
5/16	7.95	3/8	3	2	50	25	3.5
3/8	9.52	3/8	3	2 1/4	57	25	3.5
7/16	11.11	1/2	4	2 3/4	70	40	5.5
1/2	12.70	1/2	4	3	76	40	5.5
5/8	15.85	5/8	4	3 3/4	95	65	9.0
3/4	19.05	3/4	4	4 1/2	114	100	14.0
7/8	22.22	1	5	5 1/4	133	165	23.0
1	25.40	1	5	6	152	165	23.0
1 1/4	31.75	1 1/4	5	7 1/2	190	250	35.0
1 3/8	34.92	1 1/2	6	8 1/4	210	375	52.0
1 1/2	38.10	1 1/2	6	9	230	375	52.0
1 3/4	44.45	1 3/4	6	10 1/2	267	560	78.0

NOTE: The spacing of clips should be six times the diameter of the wire rope. For end-to-end connections, increase the number of clips listed above by two, and apply the proper torque listed above on all clips. Reverse U-bolts at the center of the connection, so that they are on the dead (reduced load) end of each wire rope.

HOOKS AND SHACKLES

13-12. When using hooks and shackles—

- Replace bent, twisted, or otherwise damaged hooks, shackles, rings, and pad eyes and other fittings that show excessive wear.
- Close all hooks used to support human loads or loads that pass over workers. Where there is danger of relieving the tension on the cable due to the load or hook catching or fouling (such as lowering a load in a confined space), do not use open hooks.
- Extend the dead end of the wire rope a distance of six to nine times the wire rope's diameter beyond the socket when using a wedge-socket fastener *(Figure 13-2)*. Clamp the dead end of the wire rope to an extra piece of short wire rope. Place the clamp as close to the wedge as possible. Never clamp the dead end of wire to the live end. The U-bolt should bear against the tail, and the saddle of the clip should bear against the short extra piece.

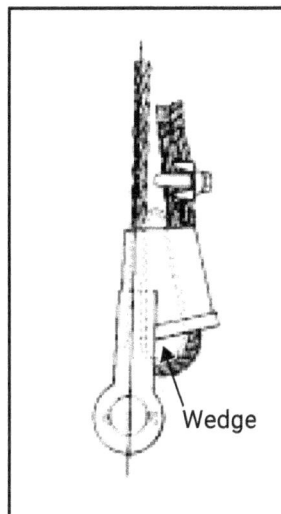

Figure 13-2. Wedge-Socket Fastener

EQUIPMENT LOADING

13-13. When loading equipment using suspended or overhead loading devices, the operator of the truck to be loaded must leave the cab. When loading equipment for transporting—

- Distribute, chock, and tie down the load.
- Make sure that the forward and side vision are clear, and that the load does not interfere in any way with the safe operation of the vehicle.
- Keep the load within the sides of the hauling equipment. When working with oversize loads that extend beyond the sides or end of hauling equipment, give adequate warnings and precautions to prevent endangering passing or opposing traffic, or damaging the vehicle.

EQUIPMENT TRANSPORTING

13-14. Give special consideration to equipment that is wider than standard vehicles or that protrudes past the sides or end of the hauling trailer. When oversize equipment is traveling or being transported from one job site to another—

- Travel or transport during daylight hours.
- Use warning flags and/or lights.
- Accompany the equipment with a lead and/or follow vehicle with warning signs whenever possible, especially in congested or heavy traffic areas.

NIGHT OPERATIONS

13-15. When operating equipment at night—

- Equip all mobile equipment with adequate headlights and taillights.
- Keep construction roads and working areas well illuminated until all workers have left the area.
- Ensure that signalers, spotters, inspectors, maintenance personnel, and others who work in dark areas exposed to vehicular traffic wear reflectorized vests or other such apparel if the tactical situation permits.

EXCAVATIONS

13-16. When excavating—

- Shore, brace, or slope excavations that are more than 4 feet deep, unless working in solid rock, hard shale, hardpan, cemented sand and gravel, or other similar materials.
- Design shoring and bracing to be effective all the way to the bottom of the excavation.
- Use sheet piling, bracing, shoring, trench boxes, or other methods of protection, including sloping, based upon calculation of the pressures exerted by and the condition and nature of the materials being retained.
- Provide additional shoring and bracing to prevent slides or cave-ins when excavating or trenching in locations adjacent to backfilled excavations or when subjected to vibrations from traffic, vehicles, or machinery.

Chapter 14

Environmental Protection

US military forces operate under increasingly diverse environmental requirements, both domestic and foreign, particularly during stability operations and support operations. Heightened environmental concern has led all federal agencies, including the armed services, to consider the environmental consequences of proposed actions. Compliance with environmental laws and regulations is now a necessary cost of doing business. The military must comply with all environmental laws and regulations that apply to installations or theaters of operation. See TC 20-401 for additional information concerning military environmental protection.

PREOPERATIONS CHECKLIST

14-1. Unit leaders must consider the environmental impact of unit actions and plan to eliminate or minimize negative effects. An example of a preoperations checklist is shown below.

- Has coordination occurred with the installation and operational staffs concerning applicable environmental laws, regulations, and considerations?
- Has the operation plan (OPLAN)/operation order (OPORD) and or constructive directives been reviewed for environmental requirements or considerations?
- Has the environmental-protection portion of the unit's SOP been reviewed, especially the areas concerning spill response and reporting?
- Have land-use permits and range clearances been obtained? Have all other unique, environmental-related requirements been completed?
- Has a recent risk assessment been performed to identify environmental-related risks?
- Have areas of environmental concern been verified during site reconnaissance?
- Are rehearsals conducted to ensure that all safety and environmental considerations are satisfied?
- Are hazardous material (HM) and petroleum, oils, and lubricants (POL) checked to ensure that they are properly labeled and that a material safety data sheet (MSDS) is on hand for each substance before transporting?
- Are personnel designated for the spill-response team(s) properly trained and aware of their assignment?

- Are tools, equipment, and materials (spill kits) available for environmental emergencies?
- Have previous after action reviews (AARs) been evaluated for environmental lessons learned concerning the area(s) of operation and the mission type, and have preventive measures been taken?

PERSONNEL-PREPARATION CHECKLIST

14-2. An example of a personnel-preparation checklist is shown below.

- Do soldiers understand their responsibilities in reducing the generation of hazardous waste (HW) and minimizing damage to the environment?
- Have all soldiers been briefed on range or maneuver restrictions and endangered species, vegetation, archeological, cultural, and historic resource preservation?
- Are identified environmental risks discussed during planning?

SPILL-RESPONSE PLAN

14-3. A spill is defined as any quantity of petroleum product over 5 gallons (or according to local laws since some states are more stringent than 5 gallons) or any quantity of HW. Should a spill occur, take the following actions immediately (*Figure 14-1*).

14-4. Protect yourself by—

- Evacuating the area in a manner appropriate to the type of spill.
- Taking personal precautions as detailed in the MSDS for the type of material spilled.
- Using the proper personal protective equipment.
- Extinguishing smoking materials and all sources of ignition.
- Turning off power if there is the possibility of fire.
- Ventilating the area.
- Showering and changing clothes (as soon as possible) if HW contamination occurs.

14-5. Stop the flow if it is possible, without taking unnecessary chances or causing injury or contamination, by shutting off valves and turning drums upright.

14-6. Contain the spill by—

- Using an absorbent.
- Making dams to keep the spill from spreading farther. Do not let it enter storm or sewer drains or other waterways.
- Diverting the flow to prevent the spill from entering any water source (including drains) if containment is not possible.

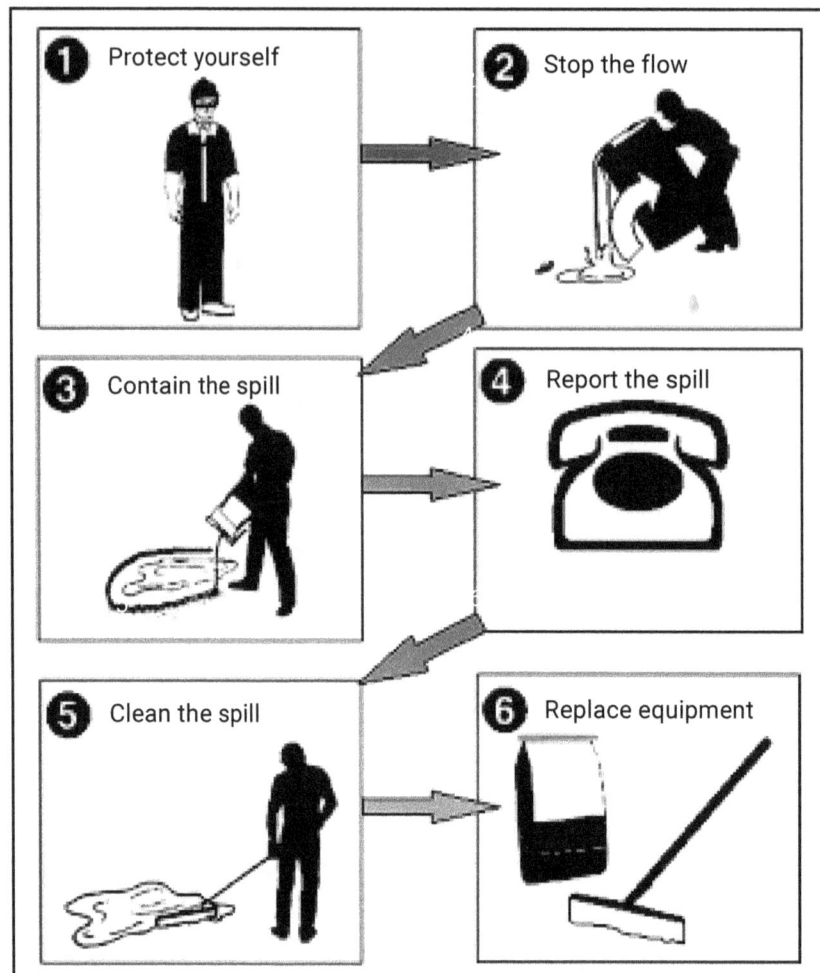

Figure 14-1. Spill-Response Actions

14-7. Report the spill to a superior and—

- Sound an alarm or give a verbal warning.
- Continue to assess the size and severity of the spill, while having another person call the installation fire department.
- Report the spill of any HM other than a petroleum product (regardless of the quantity) to the unit environmental compliance officer or the installation's environmental office immediately.

14-8. Clean the spill by—

- Scooping up the contaminated material and put it in a container. Mark the container "Hazardous Waste, Contaminated Dirt" if the spill occurred on concrete or asphalt and the spill was cleaned up with dirt.
- Checking with the unit supply sergeant or the Defense Reutilization Management Office (DRMO) for proper disposal procedures.

14-9. Replace the spill equipment through the following procedures—

- Immediately after a spill is cleaned, the spill-response team's noncommissioned officer in charge (NCOIC) will account for all tools and supplies. The NCOIC will order replacement consumables (sweeping compound and rags) from unit supply. The NCOIC will also identify missing property and initiate appropriate action (statement of charges or report of survey) to maintain accountability.
- Before resealing the drums, the spill-response team's NCOIC will complete a spill-kit inventory.

APPENDIX A

Metric Conversion Chart

This appendix complies with current Army directives, which state that the metric system will be incorporated into all new publications. The following table is an English-to-metric conversion chart.

Table A-1. Metric Conversion Chart

US Units	Multiplied By	Equals Metric Units	Metric Units	Multiplied By	Equals US Units
Length					
Inches	2.5400	Centimeters	Centimeters	0.3937	Inches
Feet	0.3048	Meters	Meters	3.2808	Feet
Yards	0.9144	Meters	Meters	1.0936	Yards
Miles	1.6093	Kilometers	Kilometers	0.6214	Miles
Area					
Square inches	6.4516	Square centimeters	Square centimeters	0.1550	Square inches
Square feet	0.0929	Square meters	Square meters	10.7640	Square feet
Square yards	0.8361	Square meters	Square meters	1.1960	Square yards
Acres	0.4047	Hectares	Hectares	2.4710	Acres
Volume					
Cubic feet	0.0283	Cubic meters	Cubic meters	35.3144	Cubic feet
Cubic yards	0.7646	Cubic meters	Cubic meters	1.3079	Cubic yards
Gallons	3.7854	Liters	Liters	0.2642	Gallons
Weight					
Foot-pounds	1.3830	Newton-meters	Newton-meters	0.7380	Foot-pounds
Pounds	0.4536	Kilograms	Kilograms	2.2046	Pounds
Foot-pounds	1.3830	Newton-meters	Newton-meters	0.7380	Foot-pounds

Glossary

AAR	after action review
AFJPAM	Air Force joint pamphlet
AFJMAN	Air Force joint manual
aggregate	a clustered mass of individual soil particles varied in shape, ranging in size from a microscopic granule to a small crumb, and considered the basic structural unit of soil
AR	Army regulation
atomize	to reduce to a fine spray
Atterberg limits	water contents at certain critical stages in soil behavior; they can be used to describe the plasticity of a soil and if the soil is cohesive or cohesionless
attn	attention
ballast	a heavy substance (such as wet sand) used to increase the weight of rollers
base course or base	important element in a road structure; it functions as the primary load-bearing component of the road, ultimately providing the pavement (or surface) strength; therefore, it is made of higher quality material than subbase material
BCY	bank cubic yard(s)
binder	a material that produces cohesion in loosely assembled substances; for example, tar, cement, and cohesive soil material passing a Number 40 sieve
borrow pit	an area where material is excavated for use as fill at another location
CCY	compacted cubic yard(s)
cfm	cubic feet per minute
chock	motionless; for blocking the movement of a wheel
clay	a cohesive soil that exhibits plasticity within a range of water contents and whose particles are less than 0.005 millimeters in size
coefficient	any of the factors of a product considered in relation to a specific factor, especially a constant factor of a term as distinguished from a variable
cohesion	the act or state of sticking together tightly
cycle time	cycle time is the time required for a machine to complete one cycle of operation
DA	Department of the Army

desired dry density	usually expressed as an acceptable density range but stated as a single value when used to determine soil stabilizing requirements
DEUCE	deployable universal combat earthmover
dredging	method of moving material from below a body of water
DRMO	Defense Reutilization Management Office
efficiency factor	a percentage factor (60-minute working hour = 100 percent) used to adjust production estimates for normal production delays
EM	engineer manual
EVW	empty vehicle weight
F	Fahrenheit
finishing	the final grading of an embankment or other earthwork or the smoothing of a wearing surface after it is placed
FM	field manual
FMTV	family of medium tactical vehicles
fpm	foot, feet per minute
FSN	federal stock number
ft	foot, feet
gantry	a triangular frame on top of a crane superstructure, which carries sheaves for the boom support lines; also, a platform (usually supported by towers) made to carry a traveling crane on parallel tracks
gap graded	*see* soil gradation
GPM	gallons(s) per minute
gradation	*see* soil gradation
granular	consisting of particles having a bulky shape
gravel	*see* soil
GVW	gross vehicle weight
heaped	material piled above the sides of a restricting container (such as an excavator bucket, a scraper bowl, or a dump-truck carrying box)
HM	hazardous material
hopper	usually, a funnel-shaped receptacle for holding and loading material (grain, sand, crushed rock, or coal); also, any of various other receptacles for the temporary storage of material
HQ	headquarters
HW	hazardous waste
in situ	soil in its natural (undisturbed) state
in-place mixing	mixing done at the construction site
inst	institute
kph	kilometers(s) per hour
lb	pound(s)

LCY	loose cubic yard(s)
lift	the depth of material that may be placed or compacted at one time
load time	the time it takes the loading equipment to actually load the haul unit, plus any time lost by the loading equipment while waiting for the haul unit to be spotted
loam	a general agricultural term applied most frequently to sandy, silty topsoils that contain a trace of clay
M-Kg	meters to kilograms
mph	mile(s) per hour
MSDS	material safety data sheet
NA	not applicable
NATO	North Atlantic Treaty Organization
NAVFAC	naval facility
NCOIC	noncommissioned officer in charge
NSN	national stock number
No.	number
OMC	optimum moisture content
OPLAN	operation plan
OPORD	operation order
optimum moisture content	the moisture content at which the soil's highest density can be obtained for a given amount of compactive input energy; soils compacted at moisture contents below optimum do not compact as completely as those at optimum moisture; those above optimum approach a plastic stage and begin to act like liquids, distributing an applied force equally in all directions and not moving particles into the voids
outriggers	stabilizers used on cranes and backhoes to prevent tipping while loading or digging
pcf	pounds per cubic foot
PCSA	Power Crane and Shovel Association
PI	plasticity index
pintle	a pivot pin (usually upright) on which another part turns
plasticity	the ability of a soil to deform without cracking or breaking; see also optimum moisture content
POL	petroleum, oils, and lubricants
psi	pound(s) per square inch
push loading	loading a scraper with dozer (push tractor) assistance
push tractor or pusher assistance	a dozer pushing a scraper during earthmoving operations
rimpull	the usable force developed between the driving tires and the travel surface

ripping	digging or tearing hard material using shanks (teeth) mounted on a dozer, grader, or other machine; the number of shanks mounted on the back of a dozer can usually be changed to engage one, two, or three shanks
ROPS	rollover protective system
rpm	revolution(s) per minute
RPR	rimpull required
SCIP	scarify and compact in place
SEE	small emplacement excavator
shore	(1) to give support to; brace; (2) a prop for preventing sinking or sagging; (3) a prop placed against or beneath equipment to restrict movement
shoulder	that part of the top surface of an approach embankment, causeway, or cut immediately adjoining the roadway that accommodates stopped vehicles in emergencies and laterally supports base and surface courses
side casting	to push or throw to the side, using with the blade or bucket
soil	soil is classified by particle size and type; gravel has large, coarse, blocky-shaped particles, while clay has small, fine, platy-shaped particles; sand and silt have particle sizes between these two extremes; (for earthmoving, soil is placed in three categories: rock, soil, and rock soil)
soil gradation	soil is either well-graded or poorly graded; well-graded soil is capable of being tightly compacted; it contains a variety of particle sizes; during compaction, smaller particles are worked between and around larger particles to reduce the percentage of voids, making the soil denser and stronger; poorly graded soil is difficult or impossible to compact; it contains a high percentage of similar-size particles (called uniformly gapped) or a poor relationship of the percentage of sizes (called gap -graded); such soil has a relatively high percentage of voids after compaction; therefore, it lacks density and strength
SOP	standing operating procedure
sq	square
STP	soldier training publication
struck	a full load of material that is level with the top of its container, (such as a scraper bowl or a dump-truck body)
tandem	a group of two or more arranged one behind the other or used or acting in conjunction
TB	technical bulletin
TC	training circular
tine	a slender, pointed projecting part; a prong
TM	technical manual

torque	a force that produces or tends to produce rotation or torsion (such as an auto engine delivers to the drive shaft)
TRADOC	United States Army Training and Doctrine Command
US	United States
USAES	United States Army Engineer School
USCS	Unified Soil Classification System
vpm	vibrations per minute
windrows	a long, low ridge of material scraped to the side, using a blade, when moving earth
yd	yard(s)

Bibliography

AR 600-55. *The Army Driver and Operator Standardization Program (Selection, Training, Testing, and Licensing).* 31 December 1993.

Caterpillar® Performance Handbook, 28th Edition. Caterpillar Incorporated, Peoria, Illinois. October 1997.

Compressed Air Handbook. Compressed Air and Gas Institute. 1947.

Construction Planning, Equipment, and Methods, 5th Edition. Robert L. Peurifoy, William D. Ledbetter, and Clifford J. Schexnayder. McGraw Hill Company, Incorporated. 1996.

EM 385-1-1. *Safety and Health Requirements Manual.* 3 September 1996.

FM 5-34. *Engineer Field Data.* 30 September 99.

FM 5-125. *Rigging Techniques, Procedures, and Applications.* 3 October 1995.

FM 5-134. *Pile Construction.* 18 April 1985.

FM 5-410. *Military Soils Engineering.* 23 December 1992.

FM 5-412. *Project Management.* 13 June 1994.

FM 5-430-00-1. *Planning and Design of Roads, Airfields, and Heliports in the Theater of Operations—Road Design.* AFJPAM 32-8013, Volume I. 26 August 1994.

FM 5-430-00-2. *Planning and Design of Roads, Airfields, and Heliports in the Theater of Operations—Airfield and Heliport Design.* AFJPAM 32-8013, Volume II. 29 September 1994.

FM 5-436. *Paving and Surfacing Operations.* To be published within six months.

FM 5-472. *Materials Testing.* NAVFAC MO 330/AFJMAN 32-1221(I). To be published within six months.

STP 5-62E12-SM-TG. *Soldier's Manual and Trainer's Guide: Military Occupational Specialty 62E, Heavy Construction Equipment Operator, Skill Levels 1/2.* 25 September 1985.

TB 43-0142. *Safety Inspection and Testing of Lifting Devices.* 28 February 1997.

TC 20-401. *The Soldier and the Environment.* To be published within six months.

TM 5-332. *Pits and Quarries.* 15 December 1967.

TM 5-2330-360-14&P. *Operator's, Organizational, Direct Support and General Support Maintenance Manual, Including Repair Parts and Special Tools List for Semitrailer, Low-bed: 40-Ton Construction Equipment Transporter, M870 (CCE) (CMI/Load King Model 403LF) (NSN 2330-00-133-1731).* 6 December 1984.

TM 5-2410-233-20. *Organizational Maintenance Manual Tractor, Full-Tracked, Low Speed; DED; Medium Drawbar Pull; Oscillating Track, 78-Inch Gage (Caterpillar Model D7F) with Ripper (NSN 2410-00-177-7283); with Ripper and ROPS (NSN 2410-00-185-9794); with Ripper, ROPS (CAB) Winterized (NSN 2410-00-30-6665); with Winch (NSN 2410-00-177-7284); with*

Winch and ROPS (NSN 2410-00-185-9792) and with Winch, ROPS (CAB) Winterized (NSN 2410-00-300-6664). 31 August 1973.

TM 5-3805-248-14&P-1. *Scraper, Earthmoving, Motorized; Diesel Engine Driven (NSN 3805-01-153-1854) Operation.* 19 August 1985.

TM 5-3805-261-10. *Operator's Manual for Grader, Heavy, Road, Motorized, Caterpillar Model 130G (NSN 3805-01-150-4795).* 30 March 1989.

TM 5-3810-232-12. *Operator and Organizational Maintenance Manual: Crane, Wheel Mounted: 20-Ton at 10-Foot Radius, 2 Engines, Diesel Engine Driven, 4 x 4 Air Transportable, 3 Phase with Blade, Bulldozer, Earthmoving with Block, Tackle, 20-Ton with Boom Crane, 30 Foot (American Hoist and Derrick Models 2380 (NSN 3810-00-763-7728) and 2385 (3810-00-043-5354).* 10 September 1970.

TM 5-3810-294-20. *Organizational Maintenance Manual for Crane-Shovel, Truck Mounted; 3/4 Cubic Yard, 20-Ton, with Clamshell, Dragline, and Backhoe Attachments, Gasoline Engine Driven. (Harnischfeger Corporation Model M32OT2) (NSN 3810-00-151-4431).* 5 April 1974.

TM 5-3825-221-15. *Operator's, Organizational, Direct Support, General Support and Depot Maintenance Manual: Distributor, Water, Tank-Type: Truck Mounted; Gasoline Driven (MacLeod Model W15A, Nonwinterized) (NSN 3825-00-954-9033); Multifuel Driven (MacLeod Model W15A, Winterized) (3825-00-774-9090); Multifuel Driven (MacLeod Model W15A4112) (3825-00-077-0550) and Diesel Driven (MacLeod Model W15E9019) (3825-00-474-3742).* 16 December 1964.

TM 5-3895-275-12. *Operator and Organizational Maintenance Manual for Paving Machine, Bituminous Material, Gasoline Engine Driven: Crawler Mounted, 8- to 10-Foot Wide (Barber-Greene Model SA-35) (NSN 3895-057-8715).* 31 December 1974.

TM 5-3895-330-10. *Operator's Manual for Spreader, Aggregate, Towed: Force Feed, Pneumatic Tires; 8-Foot Width (Burch Corporation, Model FF-8) (FSN 3895-130-3633).* 17 May 71.

TM 5-3895-334-15. *Operator, Unit, Direct Support, General Support, and Depot Maintenance Manual for Kettle, Heating, Bituminous, Gasoline Engine; Wheel Mounted, 2 Pneumatic Tires; 165-Gallon Capacity (White Manufacturing, Model F3M-1) (NSN 3895-00-442-9741).* 29 April 1970.

TM 5-3895-353-14&P. *Operator's, Organizational, Direct Support, and General Support Maintenance Manual Including Repair Parts Information and Supplementary Operating, Maintenance and Repair Parts Instructions for Roller, Vibratory, Self-Propelled (CCE) Model SP-848 (NSN 3895-01-075-2823).* 24 April 1981.

TM 5-3895-371-10. *Operator's Manual for Bituminous Distributor Body, M918 (Model D-63) (NSN 3895-01-028-4390).* 5 October 1979.

TM 55-2410-234-14. *Highway Transport Procedures: Transport of Caterpillar Model D8K-8S-8 Tractor with Ripper or Winch on M870, 40-Ton Low-bed Semitrailer Using the M920 Truck Tractor.* 10 August 1984.

TM 5-624. *Maintenance and Repair of Surface Areas.* 27 October 1995.

TM 5-822-8. *Bituminous Pavements-Standard Practice* 30 July 1987.

TM 9-2320-272-10. *Operator's Manual for Truck, 5-Ton, 6 x 6, M939, M939A1, and M939A2 Series Trucks (Diesel).* 15 August 1996.

TM 9-2320-361-10. *Operator's Manual for 2 1/2-Ton, 6 x 6, M44A2 Series Trucks (Multifuel); Truck, Cargo: 2 1/2-Ton, 6 x 6, M35A2, M35A2C, M36A2; Truck, Tank, Fuel: 2 1/2-Ton, 6 x 6, M49A2C; Truck, Tank, Water: 2 1/2-Ton, 6 x 6, M50A2, M50A3; Truck, Van, Shop: 2 1/2-Ton, 6 x 6, M109A3; Truck, Instrument Repair Shop: 2 1/2-Ton, 6 x 6, M185A3; Truck, Tractor: 2 1/2-Ton, 6 x 6, M275A2; Truck, Dump: 2 1/2-Ton, 6 x 6, M342A2; Truck, Maintenance, Pipeline Construction: 2 1/2-Ton, 6 x 6, M756A2; Truck, Maintenance, Earth Boring and Polesetting: 2 1/2-Ton, 6 x 6, M764.* 15 December 1988.

TM 9-2330-211-14&P. *Operator's, Unit, Direct Support, and General Support Maintenance Manual (Including Repair Parts and Special Tools List) for Semitrailer, Low-Bed: 25-Ton, 4-Wheel, M172A1 (NSN 2330-00-317-6448).* 14 July 1991.

TM 9-2330-294-14. *Operator's, Organizational, Direct Support, and General Support Maintenance Manual: Semitrailer, Low-Bed, Heavy Equipment Transporter, 60-Ton, M747 (NSN 2330-00-089-7265).* 5 May 1976.

Index

By Order of the Secretary of the Army:

ERIC K. SHINSEKI
General, United States Army
Chief of Staff

Official:

Joel B. Hudson

JOEL B. HUDSON
Administrative Assistant to the
Secretary of the Army
0007502

DISTRIBUTION:

Active Army, Army National Guard, and US Army Reserve: To be distributed in accordance with the initial distribution number 115166, requirements for FM 5-434.

www.ingramcontent.com/pod-product-compliance
Lightning Source LLC
Chambersburg PA
CBHW080418030426
42335CB00020B/2498